The Transition Timeline

for a local, resilient future

The Transition Movement is an increasingly global movement which seeks to inspire, catalyse and support community responses to peak oil and climate change. It is positive and solutions-focused, and is developing a diversity of tools for building resilience and happiness around the world. From awareness-raising and local food groups, to creating local currencies and developing 'Plan Bs' for their communities, Transition movements seek to embrace the end of the Oil Age as being a tremendous opportunity – the opportunity for a profound rethink of much that we have come to take for granted.

www.transitionnetwork.org

"Peak oil and climate change are two of the greatest challenges we face today; the Transition Town movement is firmly rooted in the idea that people taking action now in their communities can not only tackle these environmental threats but also, in the process of doing so, lead more fulfilling lives. It is about hope in an otherwise bleak-seeming future. Above all, it's about the power of an alternative vision for how society could be, and not waiting for government or politicians to get it right.

The Transition Timeline is designed to bring that vision to life – with stories of what communities have already achieved, with updates on the latest scientific data, and with 'maps' that highlight key landmarks on the journey towards a zero-carbon future. It's a hugely valuable manual for anyone committed to turning dreams into reality. Don't just read this book – use it to change your world."
– Caroline Lucas MEP, leader of the Green Party of England and Wales,
and co-author of *Green Alternatives to Globalisation: A Manifesto.*

"Shaun Chamberlin ties down the uncertainties about climate, energy, food, water and population, the big scene-setters of our future, with no-nonsense authority. What we get with *The Transition Timeline* is a map of the landscape we have to find a way through. Map-making is a risky business: sooner or later someone is going to use your map and come across a treacherous swamp that isn't marked. So you need to be alert to revisions and reports from travellers. But what matters is that someone has got the key characteristics of the landscape drawn out. This is what we have to make sense of – not in the distant future, but right now.

Don't set out without *The Transition Timeline*. Take a biro. Scribble updates, comments, expressions of shock and horror, notes to cheer yourself up. By the time your copy has been rained on, stained with blackberry juice, consulted, annotated, used to press and preserve a leaf of our autumnal world, you will have a good idea of where you are, and inspiration about where you are going. It is almost as good as getting there."
– David Fleming, Director of The Lean Economy Connection, and author of *Energy and the Common Purpose*

"There is obviously no single, magic bullet solution to climate change. But if I was forced to choose one – our best hope of averting the crisis – it would definitely be Transition Towns."
– Franny Armstrong, Director of *The Age of Stupid* film

"Transition has emerged as perhaps the only real model we have for addressing our current crisis – a new, if vital, format for reconsidering our future. *The Transition Timeline* strengthens a fragile form, something that might, without a trace of irony, be called one of the last, best hopes for all of us."
– Sharon Astyk, author of *Depletion and Abundance: Life on the New Home Front* and
A Nation of Farmers: Defeating the Food Crisis on American Soil

"The work of the Transition Towns movement is incredibly important."
– Ed Miliband MP, UK Secretary of State for Energy and Climate Change

"Will the future be as rosy as *The Transition Timeline* suggests it might be? Will the people of Britain and the rest of the world begin immediately to make better decisions, taking the welfare of future generations into account? The answer to both questions is probably no. Will serious repercussions of decisions already taken (regarding fossil fuel consumption and the structuring of our economy to depend on perpetual growth for its viability) come to bite us hard before we even have a chance to implement some of the excellent recommendations contained in this book? The answer to that one is certainly yes – we are already seeing dire consequences of past economic and energy decisions. Nevertheless, without a vision of what can be, there is no alternative to a future completely constrained by the past. The ideal future set forth herein is not a useless pipe-dream. There is not a single outcome described in this book that could not realistically be achieved IF we all do things beginning now that are entirely within our ability to do. So here it is: the map and timeline of how to save our world and ourselves. Whether we WILL take up these suggestions as scheduled is a question for the cynics and dreamers to debate. For us realists, the only relevant questions are: Where do we start?, and, Will you join us?"
– Richard Heinberg, Senior Fellow of the Post Carbon Institute,
and author of eight books, including *The Party's Over* and *Peak Everything*

"The next 100 months will be a very special time for humanity. On numerous fronts, the consequences of the past 150 years of industrialisation are all simultaneously coming home to roost. Even senior experts, scientists, NGOs and political leaders fail to appreciate that the most recent evidence reveals a situation more urgent than had been expected, even by those who have been following it closely for decades. *The Transition Timeline* provides an invaluable set of innovative approaches, new narratives and creative thinking tools that will prove vital in enabling us to shape a new kind of society and a new kind of economy: stable in the long term, locally resilient, but still active in a global context, rich in quality jobs, with a strong sense of purpose and reliant on indigenous, inexhaustible energy. It should be read by everyone, immediately!"
– Paul Allen, Director of the Centre for Alternative Technology,
and project director of *Zero Carbon Britain*

"Humanity is facing a once in a species crisis. We are approaching 7 billion people and appropriating an increasing percentage of the planet's net primary productivity, posing myriad and complex problems to our future and that of the planet's ecosystems. Of all the biophysical limits to continuing our current trajectory, energy surplus per capita looms large. And, as cheap, high quality per capita energy availability declines, our current cultural paradigm of competing for conspicuous consumption must end.

In this refreshingly real and hopeful book, Shaun Chamberlin lays out the many aspects of the limits to our growth, and highlights the fact that cultural change is likely our only successful path forward. Throughout, we are offered vision and hope that mobilising locally and nationally towards civic change does not represent a sacrifice of our health or happiness – indeed, Chamberlin points out that it would be a sacrifice to continue on our Business as Usual path. We undoubtedly face serious biological and biophysical constraints that our forebears did not. *The Transition Timeline* gives us a guide on how to best use science and culture in adapting to our new situation."
– Nate Hagens, Editor of *The Oil Drum*, and former vice president of the Salomon Brothers
and Lehman Brothers investment firms

"*The Transition Timeline* builds on the success of the Transition movement in galvanising community capacity and resilience to respond to climate change and peak oil. Using the 'backcasting' technique documented in Rob Hopkins' very successful *Transition Handbook*, Shaun Chamberlin paints the picture of how we got to a better world by 2027.

Chapters dealing with the basics from food and water through to health and medicine map how Britain made this transition using positive, bottom up community and cultural adaptation combined with innovative public policies and available and appropriate technologies.

While definitely focused on empowering the community rather than the policy makers, this book is much more than a folksy agenda for comfort in the crisis. It is a serious plan to reconstruct society in the light of ecological and energetic realities, informed by the best evidence about the vortex of forces influencing the global crisis. Chamberlin runs along a knife edge between the harsh realities facing the whole of humanity on the one hand, and hope and pragmatic vision on the other, outlining a pragmatic plan for a society-wide adaptation to the energy descent future. Let's see if we can run along that knife edge; we have nothing to lose."
– David Holmgren, co-originator of the Permaculture concept,
and author of *Future Scenarios: mapping the cultural implications of peak oil and climate change*

"Highly readable and well researched - this book is a hugely valuable contribution to Transition thinking. With grace and wit Shaun Chamberlin ably scopes out the combined dangers of peak oil and climate change and shows us what we can do to avoid their worst impacts. Read it and implement its wisdom if you want to help create a liveable future."
– Dr. Stephan Harding, co-ordinator of the MSc in Holistic Science at Schumacher College,
and author of *Animate Earth: Science, Intuition and Gaia*

"It's been said that pessimism is a luxury of good times; in bad times, pessimism is a death sentence. But optimism is hard to maintain when facing the very real possibility of planetary catastrophe. What's needed is a kind of hopeful realism – or, as Shaun Chamberlin puts it, *a dark optimism*.

In *The Transition Timeline*, Chamberlin offers his dark optimism in the form of a complex vision of what's to come. He imagines not just a single future, or a binary 'good tomorrow/bad tomorrow' pairing, but four scenarios set in the late 2020s, each emerging from the tension between two critical questions: can we recognise what's happening to us, and can we escape the choices and designs that have led us to this state? Chamberlin demonstrates that only an affirmative answer to both questions will allow us to avoid disaster – and that's where the story he tells starts to get good. *The Transition Timeline* isn't another climate jeremiad, but a map of the course we'll need to take over the coming decade if we are to save our planet, and ourselves.

The Transition Timeline is a book of hopeful realism, making clear that the future we want remains in our grasp – but only for a short while longer."
– Jamais Cascio, Co-founder of WorldChanging.com, Affiliate of the Institute for the Future,
Fellow of the Institute for Ethics and Emerging Technologies, and Founder of OpenTheFuture.com

The Transition Timeline
for a local, resilient future

Shaun Chamberlin

green books

in association with
the Transition Network

Contents

Contents

Dedication

This book is dedicated to my father, Roger, who helped me learn to think for myself.

Published in 2009 by Green Books Ltd,
Foxhole, Dartington, Totnes,
Devon TQ9 6EB, UK
www.greenbooks.co.uk

in association with
the Transition Network
www.transitionnetwork.org

All photographs are by Rob Hopkins, apart from
Amelia Gregory pp.13, 82, 88, 129; Simone Kay p.14; Mike Grenville pp.22, 111; Pontus Edenberg p.39;
Jessica C p.43; Kevin Walsh p.45; Frankie Wellwood p.59, Tulane Blyth pp.75, 76; Sally Stiles p.80;
Joe Bennett p.85; Rednuht p.87; Jaime Olmo p.88; Sonya Wallace p.95; Christopher Aloi p.153.
Back cover photo of author © Jonathan Helm

The text is printed on 100% recycled paper and the covers on 75% recycled material, using vegetable inks.
Printed and bound by Cambrian Printers, Aberystwyth, Wales, UK.

ISBN 978 1 900322 56 0

Acknowledgements and context

This book was originally conceived for the EDAP (Energy Descent Action Plan) teams in the various communities giving life to the Transition process. I very much hope that it proves a useful resource for them, and for the Transition Network supporting their development, who are in turn supported by the Tudor Trust.

While the themes discussed in this book have received far less attention than they might warrant, there are a small but growing number of outstanding reports which do provide a context within which *The Transition Timeline* sits. The *Zero Carbon Britain* report produced jointly by the Centre for Alternative Technology and the Public Interest Research Centre, and *Climate Code Red* by the Carbon Equity group have both provided great inspiration, and are much referenced herein. I would also draw attention to David Holmgren's excellent *Future Scenarios*, which was released while I was writing this book, and which complements it perfectly.[1]

My personal thanks to Paul Allen, Ben Brangwyn, Maria Bushra, Rachel Cashdan, Rosalie Chamberlin, Jonathan Essex, David Fleming, Doly Garcia, Marcin Gerwin, Tim Helweg-Larsen, Rob Hopkins, Britain Houchin, Victoria Hurth, Peter Lipman, Pritesh Mehta, Becci Somerville, Chris Vernon and David Wasdell for their advice and support at various stages of the book's development, to Trent for the inspiration, and to the many others who have provided insight and strengthened the project in their various ways.

Foreword

by Rob Hopkins

Author of *The Transition Handbook* and Founder of the Transition Network

The timeline of the Transition Timeline

We live in extraordinary times. Scary times. Exhilarating times. Bewildering times. Yet times so pregnant with possibilities as to be unprecedented. Everything may well be up for grabs, as we emerge blinking into a new economic and energy world that many in government and other positions of responsibility are quick to claim that no one could possibly have seen coming. Yet some did see it coming, and their insights are of great value today as we struggle for clarity. The more I look back to the opening section of 2005's Hirsch Report, the more prescient it is:

> ". . . the peaking of world oil production presents the US and the world with an unprecedented risk management problem. As peaking is approached, liquid fuel prices and price volatility will increase dramatically, and without timely mitigation, the economic, social and political costs will be unprecedented. Viable mitigation options exist on both the supply and demand sides, but to have substantial impact, they must be initiated more than a decade in advance of peaking."

Now here we are, in the world Hirsch predicted. Price volatility ($100 a barrel in January 2008,

$147 in July 2008, $40 by February 2009) is now a fact of life, and yes, the economic, social and political costs *are* unprecedented, as are the extraordinary impacts and the deep damage that cuts in demand have had on Western economies. We may well be on the edge of the first recession underpinned by a geologically imposed oil peak.

It is reminiscent of *The Wizard of Oz*. We have been picked up in a tornado and placed back down in a world that initially looks the same as it did, yet is profoundly different. The world around us bears little in common with the last time oil cost $40 a barrel. We are slowly finding our feet in this new and unfamiliar world, one where businesses around us are closing, the world is nudging its climate tipping points, and the economic situation is profoundly altered.

As Hirsch points out, mitigation, i.e. a wartime scale mobilisation to forever break our societal addiction to oil, should have started at least ten years ago. It didn't. While we now possess sufficient understanding of the nuts and bolts of what a more localised, low energy world might look like, we lack the understanding of how to get there, how we bring it into existence.

The idea for this book came out of one of those days at work when the mind starts to

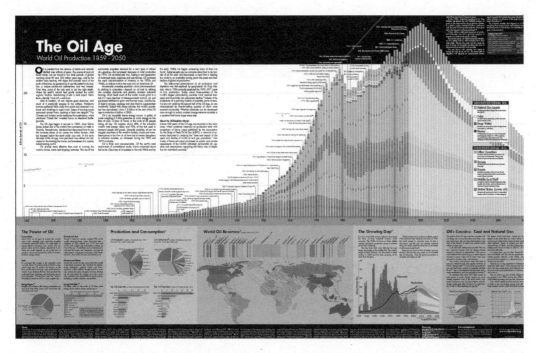

Figure 1: The Oil Age Poster, which can be ordered through http://www.oilposter.org. It is distributed free to schools and non-profit organisations.

"If a picture is worth one thousand words, then The Oil Age Poster is worth one million words . . ."
– US Congressman R. Bartlett, Maryland (Republican)

wander, around 3.30pm on a Friday afternoon, when I was looking vacantly at the now famous 'Oil Age' poster produced by the Post Carbon Institute. It offers a wonderful overview of the Oil Age, identifying the crucial points in history as humanity made the dizzying ascent to its current consumption of around 87 million barrels of oil a day.

The question I pondered, as I stared at the poster, was what the downward side, the right-hand half, of the poster might look like. I wondered whether it might be possible to start to pencil in some of the events and dates that might define our collective, careful and considered way down the mountain. In Transition Town Totnes, as part of an event we ran on Transition Tales a short while later, we spent a couple of hours coming up with

the first attempt at such a thing (see p.96). We put peak oil in 2010, the introduction of carbon rationing in 2011, 'peak cars', the point beyond which the amount of cars on the UK's roads begins its inexorable decline, in 2012. By 2015, Totnes had created its first urban market garden on the site of a former car park, 2016 saw the town introduce a free bicycle scheme and by 2017, schools had to give days off during the nut season in order to help with the harvest. By 2020, 50% of food consumed in the town was locally grown. Although our timeline was a mixture of the serious, the studied and the downright silly (e.g. 2021, first great white shark attack on the River Dart), it became an object of great fascination, people spending hours poring over it, alternately giggling and looking very serious.

"At every level the greatest obstacle to transforming the world is that we lack the clarity and imagination to conceive that it could be different."
– Roberto Unger

It struck us that there is something very powerful about such timelines, and as Transition Town Totnes entered its Energy Descent Pathways process, it rapidly became one of our key tools. Later in this book we will share some of the tools and exercises we have developed, in the hope that, combined with the thinking around Timelines developed in this book, they will be a powerful resource in planning for the future of your community.

One of the key themes of this book is stories, the ones we tell, and why we so urgently need new ones. We really only have a small handful of future stories in our culture. There is the default Business as Usual story, the one that assumes the future will be like the present, but with more of everything. Then there's the one that assumes that everything will collapse around our ears overnight, leading to a Mad Max-style world of bandits and hairy men eking out a living from mouldy potatoes and roast squirrels. Finally there is what David Holmgren calls the 'Techno-fantasy' story, the one that has us living in space stations, nipping to the Moon on holiday, growing food in bubbling tanks of chemical gloop.

As this book will set out, the first is fantasy, the second credits humanity with none of the ingenuity and creativity that got us here in the first place and the third is completely unfeasible. None of them is remotely appropriate for the first generation needing to design a successful path down from the pinnacle of the energy mountain, and to respond with sufficient purpose and depth so as to avoid runaway climate change. We need new stories, the ones about the generation who saw the problems, looked them square in the face, and responded with courage and adaptability, harnessed what excited them and acted both as midwives for the birth of a new way of living and as a hospice for the passing of the old, unsustainable way of doing things. Around the world, Transition initiatives are stepping ably into this role, and it is hoped that this book will provide them with some powerful tools to support their work.

What this book strives to do is not to set out a complete guide to creating an Energy Descent Action Plan (which will be a subsequent publication), but rather to present the context for these local plans. It arose from various Transition initiatives telling us that they found it hard to start thinking about how to design for the future evolution of their community over 20 years, as when they tried to look forward, it all looked rather foggy. So that's what Shaun Chamberlin has done so brilliantly here, to set out as clearly and eloquently as possible what that collective journey might look like. As Alvin Toffler put it,

"Our moral responsibility is not to stop the future, but to shape it . . . to channel our destiny in humane directions and to ease the trauma of transition."

Rob Hopkins
Dartington, Devon, 2009

Introduction

My personal peak oil story started in 2000, while I was studying philosophy at the University of York. Out of the blue, I received an email from my father explaining that "a long-term survey of oil and gas resources shows that demand for oil will exceed the maximum possible supply by 2010 and the oil price will sky-rocket", followed by an (enduringly plausible) outline of the likely consequences.[2]

My initial reaction, like that of so many in their 'peak oil moment', was one of shock, quickly followed by disbelief. I wondered how there could be near-universal silence on this issue if it truly had such vast implications, and tried to assure myself that 'they' would surely find some solution. Nonetheless, I resolved to look into it, and to a greater and greater extent that decision has shaped my life since.

In 2006 I met Rob Hopkins at Schumacher College and was impressed by his vision of Transition Towns. We became friends, but I still harboured doubts as to how feasible such a transition really is, given the severity of the climate and energy trends. As I became more involved with the rapidly-growing Transition movement, it quickly became clear to me that this sense of 'pessimism of the intellect, optimism of the will' was widespread among those giving their time and energy to improving the situation.

Kingston residents, including the author, launch Transition Town Kingston

Nowadays, it seems that every week there is a new report adding to the growing chorus of recognition that our society's current way of life is unsustainable. In my work, and in my wider life, I rarely meet anyone who argues otherwise. But, strangely, what seems to be less widely acknowledged is that if something is unsustainable, then, by definition, *it's going to end*.

Of course many people shy away from this conclusion because it is deeply challenging, and demands of us all that we reconsider many of our basic assumptions about our own lives and future plans, and those of our loved ones.

But once the nettle is grasped, many of us have also found that this process can be strangely inspiring and enlivening.

As Rob put it in his *Transition Handbook*:

"The question is not 'How can we keep everything going as it is?' We should instead ask how we can learn to live within realistic energy constraints. Rather than deciding our plan of action first and then picking the energy options to match it, we should start by basing our choices on asking the right questions about the energy available to underpin our plans." [3]

The question many of us were silently asking though was just how many options are really left?

This is where *The Transition Timeline* comes in. It is a first sweep at uncovering the true possibilities of our near-future, and perhaps also a balm for those who are starting to wonder whether hope is now found only in denial.

Local Transition initiatives are themselves numbered among the most hopeful signs in today's world, and this book also grew from their requests. In attempting to draft Energy Descent Action Plans (EDAPs) looking 20 years into the future of their communities, they needed to know what sort of country and what sort of world they were likely to be living in. *The Transition Timeline* helps to fill this gap, providing a sense of that wider context for EDAP teams to use in developing their plans.

Shaun Chamberlin
www.darkoptimism.org

"Any field should be judged by the degree to which it understands, anticipates, and takes action in regard to changes in society."
– Bernard Sarason (1988), *The making of an American psychologist: an autobiography*, Jossey-Bass

Finding your way around this book

On the next two pages you will find an outline summary of the latest evidence on climate change and peak oil. Cross-references are provided for those who immediately want the full detail, but this quick primer provides the key points to allow the reader to get straight into the Timeline information.

The first half of the book explores how the UK could develop against this backdrop, and is divided into three parts:

Part One considers four possible visions of our near future, and the thinking that could lead us down each path.

Part Two looks more closely at what may be considered the most desirable of these outcomes – The Transition Vision – and examines it in depth, exploring some of the key areas of concern.

Part Three, by Rob Hopkins, discusses how Transition initiatives can best use this book to support their Energy Descent Planning process.

The second half of the book may prove the most important and stimulating for some readers, contributing a number of new insights into the energy and climate challenges facing the UK and the world in the 21st century:

Part Four provides a detailed yet readable exploration of the latest evidence on climate change and peak oil, and of the critical interactions between the two.

Part Five goes on to examine their impacts in the UK, both present and future.

This book is intended as a 'living document', and is not attempting to be the final word on any of these issues. As history unfolds, new ideas, new stories and new events will surely emerge, and I hope this book will form the basis for an ongoing conversation about the future we want to create, within the Transition movement and beyond.

Climate change – a summary

The pre-industrial atmospheric CO_2 concentration was 278 parts per million (ppm) and did not vary by more than 7ppm between the years 1000 and 1800 C.E. Yet by 2005 CO_2 concentrations in our atmosphere were at 379ppm and are currently rising by between 1.5 and 3 ppm each year. By mid-2008 they had reached roughly 385ppm.[4] (p.134)

The Intergovernmental Panel on Climate Change (IPCC) reported in September 2007 that:

"if warming is not kept below 2⁰C, which will require the strongest of mitigation efforts, and currently looks very unlikely to be achieved, then substantial global impacts will occur, such as: species extinctions and millions of people at risk from drought, hunger and flooding, etc." [5]

The IPCC strictly define "very unlikely" as meaning a likelihood of less than 10%. This is because they predict 2.0-2.4°C of ultimate warming even if atmospheric CO_2 concentrations stabilised at current levels. They also state that even keeping to this level of temperature increase would involve a peak in CO_2 emissions by 2015 and 50-85% reductions in global emissions by 2050, relative to 2000 levels. (p.138 & p.139)

Moreover, there are various significant aspects to the IPCC approach which indicate that they may be *understating* the severity and urgency of the problem, with observed changes already outstripping their most pessimistic predictions. (p.140)

Drs James Hansen and Makiko Sato of NASA have found that the threshold for runaway global warming is likely to be at 1.7°C above pre-industrial levels, yet we have already seen a rise of 0.8°C, with at least an additional 0.6°C rise still due just from emissions to date. The latest science accordingly argues that we need to return atmospheric CO_2 concentrations to 300-350ppm in order to avoid catastrophe. We must reduce the already-dangerous amount of carbon in our atmosphere before temperatures increase too far and trigger feedback mechanisms. Simply reducing the rate at which our emissions continue is not sufficient. (p.146 & p.148)

Maintaining a benign climate can probably still be achieved, but to grasp this chance it will be necessary to radically and rapidly restructure our society. The years we are now living are the time when the future of our planet's climate for millennia to come will be decided.

For a fuller readable exploration of climate change, see Parts Four and Five of this book.

Peak oil – a summary

It is a fact well-established by experience that the rate of oil production (extraction) from a typical oilfield increases to a maximum point and then gradually declines. This point of maximum flow is known as the production *peak*. Because the same is true of the total oil production from a collection of oilfields the peaking concept is also applied to regions, to countries and to the entire world. This *global* production peak is what is generally referred to by the term 'peak oil'.[6] (p.116)

Global oil production has broadly levelled off at around 85-87 million barrels per day (m b/d) since mid-2005, despite the incentive to increase production caused by the massive increase in oil prices in that time (from a $13 average in 1998 to over $140 in July 2008). Many new oil wells have begun producing in this time, which means that this new production is only just managing to offset the accelerating decline in production from existing fields. (p.118)

Production losses through depletion are only going to increase around the world, and global discovery of new oil peaked back in 1965, so we are likely to see declining global production, regardless of where the oil price goes. Unfortunately, with global demand for oil projected to reach nearly 100m b/d by 2015, the price trend is likely to be upwards, with increasing numbers of people (and countries) unable to source supplies of the energy source that powers modern civilisation. (p.119)

Here in the UK our own oil and natural gas production peaked in 1999 and has been in steep decline since. Government figures forecast this production plummeting to around 15% of 1999 levels by 2027. With current trends and policies, the Government also predict that by 2010 we could be importing a third or more of the UK's annual natural gas demand. By 2020, we could be looking to import around 80% of our natural gas needs (and 75% of our coal), yet supplies are likely to face disruption, and the UK is one of the most gas-dependent countries in the world. (p.158)

The term 'peak oil' is commonly used as shorthand for energy resource depletion more generally, and the immense challenges associated with this. Due to humanity's extraordinary dependence on oil, and the lack of comparable substitutes, oil is the main focus, but other non-renewable fuels such as natural gas, coal and uranium all face depletion issues to varying degrees of urgency. When this is considered alongside the need to minimise our usage of high-carbon fuels due to climate change, it becomes clear that we must learn to live fulfilling lives using less energy. (p.127)

For a fuller readable exploration of peak oil see Parts Four and Five of this book.

Part One

CULTURAL STORIES AND VISIONS OF THE FUTURE

Chapter 1
Why cultural stories matter

On the previous two pages I outlined the trends on climate change and peak oil, which represent perhaps the most urgent and significant forces shaping our future. Yet even these challenges are, in a sense, only symptoms of an underlying reality. They are consequences of the choices we have collectively made and continue to make, and these choices are shaped by our understanding of the world – by our stories.

It is the stories that we tell ourselves about life – both individually and in our wider cultures – that allow us to make sense of the bewildering array of sensory experiences and wider evidence that we encounter. They tell us what is important, and they shape our perceptions and thoughts. This is why we use fairy stories to educate our children, why advertisers pay such extraordinary sums to present their perspectives, and why politicians present both positive and negative visions and narratives to win our votes.[7]

Totnes poet Matt Harvey telling stories at the launch of the town's EDAP process

Our cultural stories help to define who we are and they strongly impact our behaviours. One example of a dominant story in our present culture is that of 'progress' – the story that we currently live in one of the most advanced civilisations the world has ever known, and that we are advancing further and faster all the time. The definition of 'advancement' is vague – though tied in with concepts like scientific and technological progress – but the story is powerfully held. And if we hold to this cultural story then 'business as usual' is an attractive prospect – a continuation of this astonishing advancement.

"A person will worship something, have no doubt about that. We may think our tribute is paid in secret in the dark recesses of our hearts, but it will out. That which dominates our imaginations and our thoughts will determine our lives, and our character. Therefore, it behooves us to be careful what we worship, for what we are worshipping we are becoming."
– Ralph Waldo Emerson

"When people treat, say, fizzy brown sugar water as a source of their identity and human value, their resemblance to fairy-tale characters under an enchantment isn't accidental."
– John Michael Greer

"In a time of drastic change it is the learners who survive; the 'learned' find themselves fully equipped to live in a world that no longer exists."
– Eric Hoffer

"Once we lived with a sense of our own limits. We may have been a hubristic kind of animal, but we knew that our precocity was contained within a universe that was overwhelmingly beyond our influence. That sensibility is about to return. Along with it will come a sense of frustration at finding many expectations dashed."
– Richard Heinberg (2008), 'Losing Control', Post Carbon Institute

The problem with stories comes when they shape our thinking in ways that do not reflect reality and yet we refuse to change them. The evidence might support the view that this 'advanced' culture is not making us happy and is rapidly destroying our environment's ability to support us, but dominant cultural stories are powerful things, and those who challenge them tend to meet resistance and even ridicule.

The developing physical realities examined in detail in Parts Four and Five will surely change our cultural stories, whether we like it or not, but we *can* choose whether to actively engage with this process or to simply be subject to it.

The powerful cultural story that 'real change is impossible' makes it seem inevitable that current trends will continue inexorably on, yet in reality cultural stories are always shifting and changing, often subtly, but sometimes dramatically. Given their importance, then, we

should pay close attention when Sharon Astyk suggests that there are certain key historical moments at which it is possible to reshape cultural stories rapidly and dramatically, by advancing one's agenda as a logical response to events:

> *"I think it is true that had Americans been told after 9/11, 'We want you to go out and grow a victory garden and cut back on energy usage', the response would have been tremendous – it would absolutely have been possible to harness the anger and pain and frustration of those moments, and a people who desperately wanted something to do."* [8]

As Naomi Klein has argued in her book *Shock Doctrine*, this insight has until now mostly been used to advance cultural stories that benefit a few at the expense of many. Astyk contends, however, that there is no reason why, as understanding continues to spread, we could not grasp the next 'threshold moment' and build a dominant narrative linking it to the energy and climate context (to which it will almost inevitably be related), and explaining how this demands changes in our own attitudes and lifestyles. [9]

As we now look to our future, there are clearly a vast number of possibilities, but the concept of stories can help us to make some sense of it all. Here we will examine four visions of how our near-future could look, in the full awareness that the stories we tell here are themselves helping to shape the future that will come to pass.

Visions of the future – looking to 2027

	Ignoring evidence	Acknowledging challenges
Business As Usual (BAU)	*1* Denial	*2* Hitting The Wall
Cultural shift	*3* The Impossible Dream	*4* The Transition Vision

The first vision considers the continuation of the 'business as usual, things can't really be that bad' perspective that is perhaps still dominant at this time, and where it is likely to lead us. In this vision the accumulating evidence on energy resource depletion and climate change is largely ignored. I have called this vision of the future **Denial**.

Our second vision of the future explores what might happen if we collectively accept the challenging evidence emerging on resource depletion and climate change, but continue working to address it through a business-as-usual mindset. We consider what happens when 'politically realistic' actions and scientific reality collide. I have called it **Hitting The Wall**.

Our third vision documents a radical change in the cultural stories shaping our present and future. Here we see a 'cultural tipping point' as the evidence of our eyes and hearts overthrows the dominant story of 'business as usual' and replaces it with a story of taking deep satisfaction in repairing earlier mistakes, and a responsible focus on ensuring a long-term resilient future. Nonetheless, in this vision we fail to acknowledge the scale of our energy and climate challenges, meaning that while we may appear to be building a brighter future we are in essence living an **Impossible Dream**.

Our final vision of the future is the one on which we will be focusing. Here we make the same kind of cultural shift as in vision #3, but with full regard to the overwhelming urgency of the 'Peak Climate' situation. I have called this **The Transition Vision** and it will be examined in more detail in Part Two.

"If you don't know where you're going, you'll wind up someplace else."
– Yogi Berra

Chapter 2

Vision 1: Denial

Business as usual/ignoring evidence

"If you don't change direction, you are likely to end up where you're headed."
– Chinese proverb

"We have only two modes – complacency and panic."
– James R. Schlesinger, the first US Dept. of Energy secretary, on the country's approach to energy (1977)

2010
Recession deepens, UK manufacturing declared to be 'in crisis'
British Airways announces bankruptcy

World oil production passes its peak

In this possible future we failed to heed the ever-stronger evidence that we were facing a sustainability emergency until the consequences of our choices became overwhelmingly clear. As a result, we missed the opportunities to take action to prepare for the coming shocks and our fragile globalised structures were found wanting when they came.

- *Ever more desperate measures employed in the name of maintaining a growing economy*

- *Environmentally destructive energy sources like tar sands and 'coal to liquids' exploited*

- *IPCC announcement in 2019 that climate change is now unstoppable met with anger and disbelief*

The collapse of the American economy (the most indebted country in the history of the world[10]) was the beginning of the end for the story of globalisation, as its flagship foundered. By 2011 the dollar had become so devalued that it was not accepted at all in many places around the world. As it became ever more obvious that the world's greatest consumer was going to remain unable to afford the world's products a painful global economic slowdown ensued.

This was difficult enough, but the situation was not helped as over the following years it became apparent that the peak in global oil production had transpired in 2010. As the global recession deepened into depression it did reduce demand for oil, but the world's economies remained overwhelmingly dependent on abundant supplies of fossil fuels. As the realisation that the supply simply did not exist spread, oil markets panicked and the price of oil shot even higher, briefly spiking to the highest level it would ever reach (over €300 a barrel) in late 2013. Natural gas prices followed suit. In many parts of the world essential services were cut due to a lack of affordable energy, and unpredictable power cuts became routine for the majority of the world's electricity users. Those countries with available coal reserves – including the UK – increased production as fast as they were able, in desperate response to the punishing cost of energy imports and the weakening currencies of energy importers.

As the panic spread outwards from the energy markets, share prices and house prices tumbled and stock markets crashed. The ubiquitous globalised economy appeared to be dragging everyone down with it. Companies were bankrupted and mass unemployment and homelessness ensued, yet as the hole got deeper we found no better response than to dig faster.

A wave of outraged disbelief spread around the world as people began to realise that the money they had worked for decades to earn represented no security at all. Under this level of stress, cultural stories like 'every man for himself' and 'survival of the fittest' led to an increasingly desperate and individualistic struggle for the necessities of life, while wars over food, water, energy and resources sparked around the world. By the middle of the decade fascistic political parties were becoming significant political forces in many countries and environmental and political refugees in their millions were seeking new homes.

In the UK, society became ever more polarised between the 'haves' and 'have nots', with gated compounds and private security becoming the norm for the former, while the majority – and in particular groups like the elderly who were already at risk from fuel poverty – struggled to afford to keep and heat their homes.

Against this backdrop, the increasing impacts of climate change continued to assert themselves while ineffectual climate policies did little to stem the flow of emissions. Droughts affected farming yields as fish stocks collapsed due to both overfishing and the stresses of their changing environment. Water became an even more desperate issue than food in many parts of the world as droughts combined with the disappearance of snow and ice, killing the vast rivers of melt water that used to bring fresh water to billions of people and support critical ecosystems.[11]

September 22nd 2019 is the date so many of us remember. It was the day on which the Intergovernmental Panel on Climate Change (IPCC) issued its gloomy statement announcing its analysis that it was now too late to avoid unstoppable runaway climate change. The critical tipping points were passed in 2016, and climate feedbacks had now taken over from human emissions as the primary factor driving continued warming. As the news spread it was largely met with shock and anger, causing a fundamental shift in our cultural stories as the notion of leaving a world better than that we were born into seemed to become an impossibility. Desperate attempts at 'geo-engineering' appear to be only making the problems worse and we have had to face the fact that our species has probably sealed the fate of most of life on Earth.

This understanding, combined with the near-global suffering, has led to new cultural myths gaining power. Preachers speaking of God's vengeance on humanity's greed seem to be everywhere, and the political processes are increasingly dominated by empty promises of revolutionary change and protection.

View from 2027

By 2027 the world human population is in heavy decline, and it is also estimated that 30% of all species of life have become extinct, primarily through habitat loss and pollution. The rate of species loss is only increasing.

"If we had wanted to destroy as much of life on Earth as possible, there would have been no better way of doing it than to dig up and burn as much fossil hydrocarbon as we possibly could."
– Mark Lynas (2007), *Six Degrees: Our Future on a Hotter Planet, Fourth Estate*

2011
US dollar crashes, colossal losses on the international markets

House repossessions rise sharply

2012
Extreme rises in global food prices

W.H. Smith and Debenhams declare bankruptcy

London Olympics deemed 'a massive disappointment', due to the numbers of athletes and international spectators unable to attend

2013
Arctic ice-free for the first time

Major flood in Bangladesh leads to refugee crisis

London Mayor introduces free bicycle scheme to London, and bans cars from many areas of city centre

Government introduces 'Is Your Journey Really Necessary?' scheme

Oil reaches €300 a barrel, with gas prices following suit

2014
Many areas of the UK begin to experience regular power shortages

Nitrogen fertiliser becomes unaffordable for many farmers

Government presses for huge increase in coal production

2015
British Aerospace launches wind turbine manufacturing plant in Bristol

Stock-market crashes

2016
The world passes the tipping point into unstoppable climate change, although it doesn't know it at the time

2018
UK runs out of landfill sites and is forced to export virtually all of its waste

Draconian limits on rubbish collection lead to problems of waste accumulating in streets

2019
IPCC announces "it's over", and that runaway climate change is now beyond human control

2020
Huge African famine kills 4 million

60% of world's fisheries so depleted that it is 'uneconomic' to fish there

Beyond this point it becomes too unpleasant to relate

Warfare, social unrest and market chaos have also hamstrung fossil fuel consumption in recent years – global oil production is down to 22 million barrels/day and much of our fossil-fuel infrastructure lies unused and rusting.

Chapter 3

Vision 2: Hitting The Wall

Business as usual/acknowledging challenges

In this vision of the future we confronted the full reality of climate change and peak oil, but attempted to deal with them without fundamental change in our cultural stories. Acknowledgement of the challenges we face led to some action, but it was unable to be sufficiently effective within our existing frameworks.

- *Recognition of environmental challenges, but the dominant mindset states that 'there is no alternative' to business as usual*

- *Underlying trends like exponential growth of populations and economies go essentially unchallenged*

- *'Realism' about whether fundamental change in society is achievable leads to widespread despair and inaction*

As the globalised economy sank into recession and then depression, helped along by the global oil production peak in 2010, there was also a growing recognition that we were facing our last chance at maintaining a benevolent climate.

In these difficult times an ambitious global timescale – the 2010 Accord – was agreed for reducing carbon emissions to zero and beyond, yet no agreement could be reached on how to ensure that this would be achieved within an economy still exclusively striving for financial profit and growth.

Shortsighted approaches like large-scale biofuels production and nuclear energy proliferated, and in 2011 the Global Emissions Trading Scheme (GETS) arrived, with great fanfare about the sums of money that would flow through it. By this time, the food and energy markets were recognised as inextricably linked, due to the common currency of biofuels and the sheer energy-intensity of industrialised agriculture, and now carbon prices too were tied in. Some commentators spoke of a Grand Unifying Market bringing all commodities together, much as physicists were searching for a 'Theory of Everything' to unite all of physics.

When GETS launched, prices rocketed across the markets and advocates pointed to the high price of emissions permits as evidence that the market had climate change under control, especially when official measurements showed that the rate of emissions increase had slowed. Critics pointed to greed and corruption in the system (both legal and illegal) and to the hundreds of millions who

"I put a dollar in one of those change machines. Nothing changed." – George Carlin

"It may seem impossible to imagine that a technologically advanced society could choose, in essence, to destroy itself, but that is what we are now in the process of doing."
– Elizabeth Kolbert, *Field Notes from a Catastrophe*

2009
US hailed as world leader on climate issue, pushing for a strong agreement at the Copenhagen summit in December.

2010

After much wrangling and political brinksmanship, the 2010 Accord agrees to limit the peak in global atmospheric concentrations to 400ppm CO_2 and reduce from there. This is hailed as a 'victory for sanity' by environmental groups.

World reaches peak oil production

2011

Emissions already exceeding 2010 Accord targets

Global Emissions Trading Scheme (GETS) introduced

Food riots across the developing world

British Prime Minister tells President Putin "when it comes to our energy security, the UK knows who its friends are."

Oil passes €150 a barrel for first time

2012

Russia tells the UK that "in this new energy world, it is every man for himself", and brings in crippling price rises

UK experiences regular power cuts over the winter

US President Obama announces "the wheels have come off the world's oil production"

could no longer afford enough to eat. Energy supply outages also increased in severity globally and we in the UK increasingly found ourselves outbid for the small quantities of natural gas available on the export markets. By 2012 we were experiencing regular outages on the National Grid as the Government's policy of relying on imports from the global market unravelled.

It also soon became apparent that certain countries were abandoning their coal mines only to turn a blind eye as they were commandeered by illegal mining operations. This allowed them to avoid their climate obligations while also begging the international community for financial assistance in dealing with this (literal) climate crime. Meanwhile, the criminal operations dutifully paid 'black taxes' while demanding high prices from desperate energy customers.

In the light of all this suffering in both rich and poor countries, the outcry against the globalised free market became overwhelming and in 2015 GETS was abandoned in favour of a 'binding international agreement' on countries to find their own methods for reducing emissions. Some countries and organisations managed to make impressive progress, but they were hamstrung by competition with others who were neglecting their responsibilities.

Here in the UK, the brilliance of focused human ingenuity began to shine through. While the majority of the population were engaged in the struggle to support themselves and their families, those communities which had worked to develop localised food and energy supplies were faring considerably better. They were able to share their expertise, collect examples of innovative adaptation and devote time to helping other communities adapt to the post-carbon world. As international prices for fossil fuels rocketed, the cost of the UK's fossil-fuel-generated electricity also shot up, leading our Government to belatedly divert its remaining resources to supporting renewable energy and localised farming. Nonetheless, there was a pervasive sense that we would continue to pay a heavy price for the time and resources wasted in the GETS years – while certain communities were thriving in the short term, disruption of the global climate spelt an ill future for all.

View from 2027

By 2027 the world population is actually in decline and we have just passed a 455ppm concentration of CO_2 in the atmosphere. The uneven efforts around the world have had some effect, but the worldwide economic depression has done far more to reduce emissions rates. Nonetheless, atmospheric greenhouse gas concentrations are still increasing. Global liquid fuel production has dropped to 44 million barrels/day – roughly half of its peak only fifteen years earlier.

In the UK we find ourselves in a very different situation to twenty years earlier – nobody holds any faith in bank accounts or property ownership as a source of security, and just over 50% of the population are officially living in poverty. Basic infrastructure maintenance, such as road repairs, is poor. Times are hard, but we are scraping by. Unfortunately the same cannot be said of much of the rest of the world.

2013
Government introduces mandatory 'smart' electrical goods, which allow power companies to reduce the energy provided to appliances at peak times

Some countries with agreed strict limits on fossil-fuel extraction report that they are struggling in the fight against organised criminal mining projects

2014
UK fuel poverty soars

TV viewing drops sharply as the trend for home story-telling and music-making booms, a side effect of the previous winter's power cuts

2015
GETS abandoned

UK Government introduces new rules to encourage decentralised energy generation, and rushes through its 'Local Energy Act'

2016
UK Government's 'Dig for Resilience' initiative, a part of its National Food Security Plan, pays people to stay home and garden, and introduces a Lawn Tax

Chapter 4

Vision 3: The Impossible Dream
Cultural shift/ignoring evidence

"If we succumb to a dream world, then we'll wake up to a nightmare."
– Jimmy Carter, former US president

"You can't solve today's energy problems with tomorrow's new technologies."
– Jeddah Conference attendee

2009
UK Government introduces feed-in tariffs, guaranteeing a good price for households and businesses selling surplus electricity into the National Grid.

Copenhagen Climate Agreement commits all major world governments to stabilising CO_2 concentrations at 500ppm

The visions we have examined so far have largely been of the type we often hear from environmentalists – miserable and doom-laden. The Impossible Dream begins far more appealingly – but ultimately suffers the consequences of failing to acknowledge scientific reality.

- *Dominant story that 'renewable energy technology will solve climate change and peak oil for us'*

- *A lack of recognition of the full scale of the problem*

- *Minimal reduction in energy usage due to widespread expectation that a new energy source will emerge 'just in time'*

In the early years of the 21st century we saw the awareness of climate change starting to go mainstream, led by certain courageous scientists and journalists who worked tirelessly to awaken the public to this great threat. By the end of the first decade of the century this had snowballed into a mass movement, with demonstrations, camps, petitions and direct action across the world leaving politicians in no doubt that failure to address this situation was simply not an option.

In response to this pressure global governments agreed in 2009 to replace the ineffective Kyoto Protocol with binding emissions reduction trajectories for each country.

It was also recognised that the usual political excuses for missing targets could have devastating consequences in this case, so in addition, the agreement recommended that each country adopt a TEQs (Tradable Energy Quotas) energy rationing system in order to guarantee that their emissions targets would be met. This was initially taken up by only three of the signatories (including the UK), but as the system began to stimulate low carbon innovation and give those countries a head start in the new lower-energy economy, others soon followed. With global oil supplies declining from 2010 onwards, a number of countries had no choice but to ration energy anyway. By 2013 all the major emitters were operating a TEQs-style system to implement their carbon budget and fairly distribute available energy supplies.[12]

In line with the scientific advice of the IPCC the global budget was set to stabilise CO_2 concentrations at 500ppm, in order to give a significant chance of limiting global mean temperature rise to two degrees.

In the UK the carbon budget was not too onerous, and TEQs in fact proved very popular. Not only did people like the sense that they were 'doing their bit' in a functioning international agreement to defuse the threat of climate change, but when the UK TEQs scheme was launched in 2010, the majority of people found that they were below average energy users and so the scheme meant more money in their pocket each week.

Meanwhile, the Government auctioned 60% of the units to organisations to cover their energy use, and the money generated by this (around £6 billion each year) was made open to applications from communities who had come up with innovative ways to reduce their energy use. This 'Transition Fund' stimulated an explosion of Transition initiatives around the country as communities realised that the Government would support their local moves toward increased resilience and self-sufficiency in the context of the global economic slowdown.

As the globalisation project unravelled for all to see, TEQs stimulated and supported the new surge towards relocalisation, as well as huge increases in renewable energy capacity across the world. As communities globally rebuilt their resilience and self-sufficiency, demand for liquid fuels dropped, but not as quickly as supply fell away, meaning that energy prices continued to rise. In the gas-dependent UK this was especially painful, as natural gas imports became ever more expensive, and the Government's plan to rely on imports to keep the National Grid functioning was shown to be somewhat short-sighted.

As the number of families in fuel poverty steadily increased, the Government told us all that at least we were doing our bit for the global environment. Unfortunately, the elephant in the room was that the climate science picture had quietly been getting worse and worse. As more and more climate feedback mechanisms kicked in, we were starting to realise that the very notion of a 'safe CO_2 concentration level' was a flawed one, as warming became increasingly driven by other factors (see pp.142-3).

2010
World oil production passes its peak

Russia tells UK and other nations that increases in gas production are "not possible"

UK Government introduces Tradable Energy Quotas and uses the 'Transition Fund' the scheme generates to support relocalisation initiatives

New Government 'Vision for UK Food' launched, including controls on importing foods which can be produced in the UK

2011
1,000th UK Transition initiative unleashed

2012
Government introduces the 'Community Energy Act' to encourage local ownership of power generation

2013
All Copenhagen signatories adopt Tradable Energy Quotas

2016
85% of food consumed in UK grown here

2017
Global oil demand down to 70million barrels/day

2020
15% of UK energy demand met by renewables

First Climate Tribunal convened to assess possible past 'crimes against humanity'

2025
IPCC Seventh Assessment Report (AR7) expresses alarm at temperatures rising faster than expected – calls for urgent re-evaluation of agreed concentration targets

"A lie may take care of the present, but it has no future."
– Anonymous

View from 2027

By 2027, the stimulus to investment in alternatives has kick-started our move away from fossil-fuel dependency, and global demand for liquid fuels has dropped to 40 million barrels/day. We are still using all the oil and natural gas we can get our hands on, but use of dirtier fuels has been minimised as alternatives come on stream.

We have just passed 435ppm concentrations of CO_2 in the atmosphere, and are on course to our stabilisation goal of 500ppm. Unfortunately, the IPCC's latest report is telling us that global temperature is rising considerably faster than their 2007 report had predicted for these greenhouse gas levels. We are facing up to the possibility that for all we have achieved it may not have been enough, and that we now face an even greater challenge to retain a hospitable climate.

Chapter 5

Vision 4: The Transition Vision
Cultural shift/acknowledging challenges

The Transition Vision is of a resilient, more localised society in which we faced the consequences of our former cultural stories honestly, heard the lessons they brought us and moved into a thriving, satisfying future.[13]

- *Acceptance of scientific evidence leads to fundamental changes in cultural assumptions*

- *Dominant sense that 'there must be another way to live', given the unacceptable consequences of business as usual*

- *New social and economic models developed to fit new paradigm*

- *Recognition of the need to drastically and sustainably reduce energy consumption*

- *Widespread sense of hope, determination and common purpose*

As in the Impossible Dream vision (preceding pages), pressure from scientists and citizens made inevitable a global agreement to replace the ineffective Kyoto Protocol. The '2010 Accord' implemented a framework of binding emissions reduction trajectories for each country, and the UK adopted the TEQs energy

rationing system alongside it to guarantee achievement of its agreed emissions targets.[14]

Unlike in the Impossible Dream vision, the targets in the 2010 Accord were based on the latest scientific evidence and, in recognition of the consequences of failure, on the precautionary principle. The global agreement dictated that emissions must be reined in so that atmospheric concentrations never break through 400ppm CO_2, and that this must be reduced back down towards 325ppm as quickly as possible. By 2010 CO_2 concentrations had already reached 390ppm, and so the 2010 Accord mandated a global emissions peak by 2014.

Over the next couple of years a rapid shift in the public's attitude to climate change took place as a number of the foundations of denial were swept away. With a binding and sufficient global agreement in place we could no longer argue that whatever we did would be overwhelmed by the irresponsibility of other countries, and with our TEQs system in place it became transparently in everyone's personal interest to reduce their energy usage and demand better options for public transport, local food production and so forth.

"There is nothing more powerful than an idea whose time has come."
– Victor Hugo

"I awake each morning torn between a desire to save the world and a desire to savour the world. This makes it hard to plan my day."
– E.B. White

2010
EU introduces Common Sustainable Food Policy

The 2010 Accord commits all world governments to preventing atmospheric CO_2 concentrations topping 400ppm

Perhaps most crucially, as the tide of opinion and action shifted, it actually became easier to go with the flow and engage with the global common purpose than to hold out in denial. The nagging sense of a problem too huge to grapple with was replaced by the satisfaction of working together to improve the world's future.

This was also seen in international politics, as those countries that had not implemented a means of guaranteeing that their 2010 Accord commitments would be met came under intense pressure. Their arguments that they needed 'safety valves' in case they failed to meet their targets were politically unsupportable in a world where others were fulfilling their end of the bargain.

'Failure is not an option' was the unanswerable argument, and so by 2013 all the major emitters had adopted TEQs-style rationing systems to guarantee their carbon budgets and fairly distribute available energy supplies. With a shared sense of global resolve, international agreement was also reached on phasing out coal extraction altogether, pending the increasingly unlikely-looking development of reliable, energy-efficient 'carbon capture and storage'.

With the political battle for sanity won, however, the real challenges began – many countries had already begun token reductions in certain emissions but now it was time to get real.

In the UK, as elsewhere, this translated into a rapidly tightening carbon budget, with limited numbers of energy rations issued under the TEQs scheme. This made many energy-profligate habits prohibitively expensive, and inevitably led to a spate of negative opinion pieces in certain newspapers. Nonetheless, as deeper understanding of the climate crisis spread (helped by both a Government education drive and simple observation) the dominant sentiment was that unchecked climate change posed a far graver threat to our lifestyles than did the energy rationing central to the global response. In addition, rising energy prices, and limited supplies of natural gas for electricity production in particular, made energy rationing seem obviously preferable to the alternative of 'rationing by price'.

Despite the step change in the human response, however, the 'inertia' in the climate system meant that many areas of the world were still becoming steadily less hospitable as the effects of our earlier emissions made themselves felt. By 2020 millions of environmental migrants and refugees from resource conflicts were arriving in Europe, with the UK receiving its share.

Fortunately, a number of cities had foreseen this and declared themselves 'Cities of Sanctuary', as part of a campaign fronted by a number of celebrities who had themselves been refugees. These communities were determined to provide a culture of hospitality for those who had lost their homes, and worked together to provide a warm welcome, induction programmes, English lessons etc. They quickly found that they were rewarded with a new and culturally diverse set of friends and colleagues, who were able to put their skills and practical expertise to work, and indeed boost the Great Reskilling effort in those areas.[15]

As the 'feel' of our culture began shifting towards a sense of 'we're all in this together', many of the problems that seemed to be growing inexorably at the turn of the century began to recede. In 2006, 87% of Britons agreed with the statement that society today is 'too materialistic, with too much emphasis on money and not enough on the things that really matter'.[16] By 2020 that had fallen to 46%. As people spent more time living and working co-operatively in their local communities loneliness, anxiety disorders and depression declined, and the energy and application of young people was engaged with and welcomed as the different generations learnt to appreciate and utilise each other's ingenuity, skills and strengths. In the political world the Green Party also gained ever more influence as their ideas started to resonate with the current Zeitgeist.

At Government level TEQs became the centrepiece as a very different economic system rose from the ashes of the globalisation experiment. Production and consumption became seen as subordinate to quality of life, and resource use was minimised. The Happy Planet Index (HPI) was widely used internationally to estimate how efficiently and sustainably we were using our endowment of natural resources to deliver happy and long human lives.[17]

Local currencies and local investment thrived, and as agriculture developed to become less energy-intensive it became a major employment growth sector, as did jobs in green retrofitting of buildings and infrastructure, renewable energy design and installation and salvage/recycling. Much-sought-after apprenticeship schemes were set up for retraining in these areas.

Meanwhile, those who still tried to flaunt financial riches by driving a car everywhere and buying showy possessions (known as 'Clarksons') increasingly became seen at best as rather selfish and passé figures of fun. People began to laugh at their own foolish former ways and comments like 'the problem with the rat race is that even if you win, you're still a rat' became part of common folklore. Values like contentment, goodnaturedness and humility became far more highly prized, and indeed conspicuous consumption of material goods became seen as rather unacceptable at a time when everyone was pulling together to address the sustainability emergency.

A growing and voracious consumption of the pleasures of good company and time spent in nature started to develop, however, and as people's quality of life improved, the level of interest in economic measures of affluence declined. The sense of the birthing of a new, better world spread widely, in a way not seen in this country since the late 1960s.

Global agreement reached to phase out coal production

2014
UK population begins gentle reduction (around 0.3% per year)

2015
The most popular money lenders in the country are now credit unions, rather than banks and building societies

Programme introduced to reopen many of the UK railways closed in the 1960s

UK building regulations revised to ensure new buildings use 60% local materials, leading to a boom in local businesses providing materials for construction

Time Magazine declares 2015 'the Year of the Raised Bed'

2016
'Peak cars' on UK roads passed

2017
UK now imports only 10% of its food and 35% of its fruit

British car industry in tatters, many car plants converted to wind and tidal turbine manufacture

2018
Provision of car parking space by Local Authorities falls by 40%

2020
Green Party wins 40 seats in election

Obesity has fallen 40% from 2010 levels; average UK life expectancy officially reaches 81

31% of UK electricity demand met by renewable sources, partly thanks to substantial reductions in demand

2024
Pension age for both men and women raised to 66

2026
Mental health problems have reduced to directly affect just over 1 in 6 people's lives, compared with around 1 in 4 in 2007

2027
UK oil and gas production down to one-sixth of 2007 levels

70% of UK electricity is renewable

Britain 96% self-sufficient in food

Average UK life expectancy officially reaches 82, thanks to major reductions in heart disease, obesity, cancer and traffic accidents

View from 2027

By 2027 we are seeing significant developments of 'carbon drawdown' techniques, especially in agriculture, and their total effect is approaching the level of the remaining ongoing global emissions. In other words, we are nearing zero net emissions, but there is still a long way to go. Global CO_2 concentrations are at virtually 400ppm, and the science is ever clearer that this needs to return to around 325ppm in order to avert the worst of climate change. Fortunately the delay in warming caused by the sheer mass of our planet gives us time to bring concentrations back down before global temperature catches up with the effects of current atmospheric greenhouse gas levels (see p.130).

As in the Impossible Dream vision, alternative energy technologies (now informally known as 'cool tech') have grown exponentially, and we are glad to be moving away from fossil-fuel dependency. Global demand for liquid fuels has dropped to 35 million barrels/day and is falling roughly in line with oil and natural gas depletion while world population is stabilising as improved levels of education and well-being for women worldwide lead to lower birth rates. While climate change is still causing many problems, habitat restoration for many species is under way, and the number of endangered species is reducing.

Transition Bristol
Peak Oil
Can you imagine Bristol after oil?
ECONOMY
THE TIME IS NOW
Transition Bristol
Join the mailing list
Transitio

Chapter 6

Our choices

"As for the future, your task is not to foresee it, but to enable it."
– Antoine de Saint-Exupéry (1948)

"Knowing many stories is wisdom. Knowing no stories is ignorance. Knowing only one story is death."
– John Michael Greer, The Archdruid Report (blog), May 24th 2006

Perhaps the most obvious point to make from Part One is how very undesirable the 'Business As Usual' scenarios look as they come to terms with physical reality. We will examine the current trends in more detail in Part Two, but perhaps the most concise summation was that of Fred Bodsworth:

> *"If we go on as we are, we will destroy in the next century everything that the poets have been singing about for the past two thousand years."*

When discussing visions of the future, there is often a temptation to compare them with the present and decide which we prefer, but it is important to remember that this is not a valid comparison – for better or for worse we know that we cannot freeze time. Our choice is between the different 2027s we have just considered, but we do, collectively, get to choose which of them come to pass. To put it another way, we will get the future we deserve, and if we want to play an active role in that decision we must get involved, not just leave it to 'the experts', or to government.

Yet we must also remain humble, remembering the words of the wonderful Chinese proverb:

> *"When men speak of the future, the Gods laugh."*

One thing we can be sure of is that all of our stories and forecasts about the future will, to some degree, be wrong. Certainly by considering the biggest factors that are likely to shape the future we can aim to be more accurate than we might otherwise be, but we can also be led by resilience and favour those stories and systems that will be useful to us across the widest set of possible outcomes. Resilience is humility put into practice, and as we choose resilient stories and systems we make the least desirable possible futures less likely to ever unfold.

We must also be wary of becoming trapped by our stories, and remain open to changing them as surprising events take place or new evidence emerges. Ignoring the evidence or trying to manipulate it to fit our existing stories is a big part of what led us to the worst of where we are today.

Our stories of the future are not just scenarios – they are interactive, just as we are. They must change in response to the world around them just as they themselves shape its future. It is not 'simply human nature' to continue with business as usual as the world shifts around us, any more than it is 'simply human nature' to do whatever it takes to ensure a desirable collective future. Human brains do have certain tendencies, but humans have done things beautiful and terrible, mighty and pitiful, because human nature is the ability to choose our path.

With all this in mind, let us move on to examine the Transition Vision in greater detail. Through my research for this book I have faced some challenging understandings, but overall I feel encouraged that we can still choose to make the transition to a sustainable thriving future, and motivated to play my part.

In the biggest sense we are telling the story of humanity and much of life on Earth here. Let's not write the final chapter just yet.

(Right) The ending of the story of humanity is not yet written.

Part Two

A DEEPER LOOK AT THE TRANSITION VISION

– a timeline of hope

Central to the Transition movement is the development and demonstration of a clear, positive and desirable vision of our future, but in this case we have to admit that there is some truth to the old joke:

'If that's where you're headed, I wouldn't start from here.'

Clearly that is the one matter on which we have little choice, yet we should always carry with us the understanding that the present position and trends set the direction of departure for our journey, not its final destination.

The Transition Vision itself is, of course, a living story. It is developing as the practitioners and storytellers in our communities get to work, as time marches on, and as we all write new stories through our actions.

The context of climate change and peak oil (discussed in detail in Parts Four and Five) means that, whichever way we look at it, we must learn to live with less energy. Since energy is perhaps most simply defined as *the ability to do work* it is easy to see why this has wide-reaching implications, yet the stories we choose – the visions we hold – *create possibilities*.

Accordingly, in Part Two we will examine the Transition Vision in greater depth and examine what is really possible, starting from where we are. In the process we will produce an evolving overview briefing on the areas of key concern in our energy descent.

"To be truly radical is to make hope possible, rather than despair convincing."
– Raymond Williams

"Can we rely on it that a 'turning around' will be accomplished by enough people quickly enough to save the modern world? This question is often asked, but whatever answer is given to it will mislead. The answer 'yes' would lead to complacency; the answer 'no' to despair. It is desirable to leave these perplexities behind us and get down to work."
– E.F. Schumacher, *Small is Beautiful*

Chapter 7

Population and demographics

"We feel you don't have a conservation policy unless you have a population policy."
– David Brower, Founder of Friends of the Earth

"The raging monster upon the land is population growth. In its presence, sustainability is but a fragile theoretical construct."
– E.O. Wilson

Present position and trends

Population

We will start with a look at the population and demographics of the UK, as population is a multiplier for many of the other human environmental impacts we will be considering. It is first worth highlighting the startling global human population trend over the past few centuries.

Astonishingly, when John F. Kennedy was assassinated in 1963 there were fewer than half as many humans on our planet as there are in 2009.[18]

Examining the UK situation is complicated by the fact that the UK has only existed in its present form (England, Wales, Scotland and Northern Ireland) since 1922, and that data sources are scarce. In fact even modern census data may have a rather wider

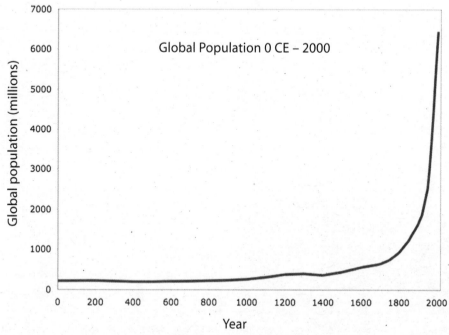

Figure 2: Growth of human population since the year 0 C.E. [19]

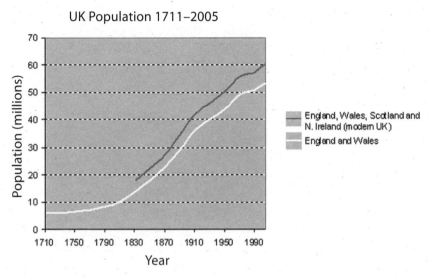

UK Population 1711–2005

Legend:
England, Wales, Scotland and N. Ireland (modern UK)
England and Wales

Figure 3: UK Population trends. **Data source at Gaia Watch UK: http://tinyurl.com/5m8xut**

margin of error than is widely appreciated. Nonetheless, the broad trends are clear, and Figure 3 gives a good indication of UK population growth during the period of this global population surge. The UK in 2009 is one of the most densely populated countries in Europe.[20]

While the world population is a simple product of birth rates and death rates, national population is also dependent on emigration and immigration. In 1998 net immigration overtook natural change (the simple effect of births and deaths) as the main driver of population growth in the United Kingdom, and despite a slight reduction in net immigration since 2004 it remains the larger contributor, due to a number of factors including the enlargement of the EU. We may also be starting to see the first trickle of the coming flood of 'environmental migrants'.

As we can see from Figure 3, it appeared that UK population growth was slowing as a result of the 'demographic transition', but over recent years the rate of increase has started rising again and the UK population is now growing

"The UN predicts that there will be millions of environmental migrants by 2020 with climate change as one of the major drivers of this phenomenon... Europe must expect substantially increased migratory pressure." – 'Climate Change and International Security: Paper from the High Representative and the European Commission to the European Council' (2008)

at the fastest rate since the 1960s. This is caused not only by migration but also increasing life expectancy and a steady rise in the total fertility rate from 1.63 children per woman in 2001 to 1.9 children per woman in 2007.[21]

It remains to be seen whether this recent increase in growth rate will be sustained, but according to the UK Statistics Authority our population is projected to rise from 61.4m in 2008 to 69.6m by 2027, thus adding 8.2m people over the period we are examining – more than the present population of London (around 7.5m).[22]

Age demographics

Other key aspects of UK demographics can be most clearly illustrated via a 'population pyramid' (Figure 4) showing the number of people of different ages in the country. In a country with a very fast-growing population the base of the pyramid would be much broader than the top, but in a relatively slow-growing population like ours there is less difference in width between the base and the top of the diagram.

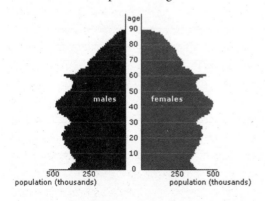

Figure 4: 2007 UK Population Pyramid.[23]
Source: National Statistics website.

At first glance, the pyramid may not seem particularly revelatory, but as we look more closely its shape allows us to see the major trends in UK population over the last century. The greater number of women relative to men at the top of the diagram reflects not only women's higher life expectancy, but also the number of men who died during the Second World War. Moving down the diagram we can see that women born during the peak years immediately after World War Two have now reached retirement age (at 60 years). Men born during this period will reach retirement age in 2012 (at age 65 years).

Further down we see the 'baby boom' of the 1960s (ages 38–48), the low fertility rates of the mid to late 1970s (ages 30–36) and the late 1980s and 1990s (ages 5–19), and the recent fertility rate increases (ages 0–5).

In mid-2007 the median average age in the UK was 39 years, up from 37 in 1997. Within this, children aged under sixteen represented around 20% of the total population, as did those of retirement age. People of working age represented 62% of the population, with around half of these over 40 years old.

The challenge is that our population continues to grow older, and by 2027 the 'baby boom' generation will be reaching the current retirement age, thus significantly increasing the proportion of our population outside of working age, who rely on services and pensions that must be maintained by those of working age. In 2006, there were 3.3 people of working age for every person of retirement age. This is projected to fall to 2.9 by 2031.[24]

Urban-rural balance

The other significant demographic trends in the UK over the past 50 years have been what is termed the 'counter-urbanisation cascade' – city dwellers leaving dense urban areas for relatively more rural settings – and the migration of people from the north to the south of Britain (which was strongest in the late 1970s and throughout the 1980s, and may actually have reversed over the last few years).[25]

Contrary to the popular conception, the south-east region has seen a net outflow of domestic migration, but this has been more than compensated by the proportion of international migration that flows to this area, with this region accordingly seeing the highest increases in population density.[26]

Partly in response to this, Greater London has sprawled out further and further, with the M25 becoming its 'inner ring road' and its commuter belt extending up to the ends of the M3 and M11, up to Leamington Spa on the M40 and to Chepstow on the M4. By some estimates, half the population of the UK now lives within the immediate influence of Greater London.[27]

We will consider the impacts of this further in the chapter on food and water (from p.49).

Cultural story change

The vast expansion in human numbers on our planet is often presented as a triumphant symbol of human ingenuity and progress, yet the evidence suggests a different perspective. In September 2007 Dr Stuart Pimm of Columbia University told the United Nations that:

> "The entire web of life is on the verge of catastrophe . . . If things continue, in as little as 35 years half of all species of life will be extinct." [28]

This web of life includes humanity. Pimm warned that "billions of people could die because of decreased biodiversity", but this is not the bottom line. Nature is a complex interdependent system of which we are one significant part, and if Nature becomes so unbalanced as to see super-exponential increases in human population accompanied by collapses almost everywhere else in the system, then the system can begin to break down, with lethal results for the entire web of life.

We are entirely dependent on Nature's processes for our food, drinking water and even oxygen, yet there is a powerful cultural story of humanity's separation from (and indeed domination over) Nature, which has led to our complacency in the face of the mass extinction taking place as the consequence of our choices and actions.[29]

With regard to these stresses caused by ever-expanding population, economist

Kenneth Boulding produced what he calls his 'Dismal Theorem':

> *"If the only ultimate check on the growth of population is misery, then the population will grow until it is miserable enough to stop its growth."*

which he followed with his 'Utterly Dismal Theorem':

> *"Any technical improvement can only relieve misery for a while, for so long as misery is the only check on population, the [technical] improvement will enable population to grow, and will soon enable more people to live in misery than before.*
>
> *The final result of [technical] improvements, therefore, is to increase the equilibrium population which is to increase the total sum of human misery."*

Fortunately he concluded with 'The moderately cheerful form of the Dismal Theorem':

> *"If something else, other than misery and starvation, can be found which will keep a prosperous population in check, the population does not have to grow until it is miserable and starves, and it can be stably prosperous."* [30]

Our aim, then, must be to find means of stabilising population other than misery, lest we find ourselves working to spread ever-dwindling resources among ever-growing populations. This is why, despite the UK demographic challenges we have considered, many scientists urge us to voluntarily limit our family sizes. The strategy of having more children in order to prop up our pension system would commit us to the dismal prospect of indefinite population growth, as today's babies become tomorrow's pensioners. [31]

Perhaps the change in perspective we need here is to stop seeing elders as essentially useless burdens on our working-age population. In many cultures the elders are the most highly-valued and respected members of the community, with a number of crucial roles such as caring for grandchildren and great-grandchildren, acting as the repositories of knowledge and wisdom and indeed often being among the key decision-makers. In our present time and culture their memories of lower-energy lifestyles and the practical skills that they require are likely to be especially pertinent.

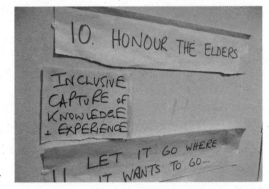

Transition principles

We also need to challenge the pervasive notion that population growth is essentially a developing-world problem, with UK contributions almost irrelevant. As the Optimum Population Trust highlight:

"each new UK birth will be responsible for 35 times more greenhouse gas emissions and associated environmental damage than a new birth in Bangladesh and 160 times more than a birth in Ethiopia." [32]

While population growth rates are significantly higher in Bangladesh (2%) and Ethiopia (3.2%) than in the UK (0.3%) the difference between our levels of consumption is an order of magnitude greater, meaning that the UK growth rate in greenhouse gas emissions and ecological impact due to increasing population is considerably higher than that of these developing countries.[33]

This also emphasises that population is not (as some claim) the single most crucial environmental issue. It clearly has a significant effect as a multiplier, but our chosen way of life and ecological footprint are bigger contributors to climate change, energy resource depletion and the other challenges facing us today and in the near future.

Perhaps the final set of cultural stories that is relevant here concerns our relationship with and responsibilities to future generations. Here in the UK the dominant ethic among many has been to strive to see your children better off than you yourself were, yet we also encounter the conflicting notion of 'why should I care about that? I'll be dead by then.'

To use Einstein's concept (see sidebar opposite), we need to widen the circle of our compassion not just to all other humans and the whole of Nature, but also through time, to encompass those who will come after us.

With all these possible cultural shifts in mind, we can return to Dr Stuart Pimm's blunt yet hopeful conclusion to that address to the United Nations Climate Change Conference:

"When the time is right, society can change. We are either going to wake up or die – we don't know which it will be." [34]

Let's work on the assumption that the time is indeed right for us to change and 'wake up', and examine how the demographics of the UK could develop over the next twenty years as we collectively embrace these compassionate stories and let them guide our actions.

The Transition Vision – looking back from 2027

As awareness of the growing ecological crisis spread, the question of sustainable human population levels started to appear in the media with ever-increasing frequency throughout 2009 and 2010. This growing profile helped to move the question of population into the public eye, and allowed politicians and other public figures to begin discussing this once taboo subject. Fertility treatment in particular became controversial, with a growing sense that increasing fertility might not be the best use for the considerable energy and resources involved, and many couples instead opting to contribute to the rise in adoptions.

Yet the real breakthrough came when the soaps got involved. In 2010 both *Coronation Street* and *EastEnders* introduced long-running storylines in which characters

"You do not inherit the Earth from your ancestors, you borrow it from your children."
– Traditional

grappled with the moral implications of having more children. These storylines were widely discussed and analysed in both the media and living-rooms around the country, and in 2011 we saw a significant and sustained drop in birth rates, with no other obvious cause.

Wanting to keep up with this shift in cultural attitudes, our Governments started legislating accordingly, structuring taxes and benefits to encourage smaller family sizes and changing their immigration criteria – offering citizenship to more refugees (including climate refugees) but to far fewer economic migrants, and aiming for zero net migration.

The outcome was that the UK population peaked in 2013, and since then has been slowly reducing (at around 0.3% per year) to our current level of 58.9 million in 2027.

As communities have tended to become more resilient and self-sufficient, there has also been a significant decrease in internal migration over the past 20 years, with people far more likely to live and work in the area where they were born, meaning that people tend to live much closer to their extended family. This provides an important source of support to many and has also led to a reduction in so-called 'love miles' – the travel required for visiting distant relatives.

As predicted we now have only three people of what used to be termed 'working age' for each person over 66 years old, and this figure is set to decrease due to the dropping proportion of under-16s in the population. As announced in 2006, the pension age for both men and women rose to 66 in 2024, and it is set to rise again in the coming decade.[35] However, with flexible employment opportunities being offered, and a greater recognition of their value in a strong extended family, older people have often found satisfaction in being seen as crucial contributors to the transition of UK society.

Chapter 8

Food and water

Present position and trends

Along with air, water and food are the most fundamental requirements of human life, so ensuring a reliable healthy supply is one of the first questions we must examine when looking to the future.

Global food trends

Until recently, the UK had benefited from over 50 years of declining food prices, partly thanks to the food and farming policy created after the Second World War to encourage indigenous production. Now, however, with a 12% decline in UK indigenous food production over the last ten years and our Government deciding to rely increasingly on purchasing imported food, we are seeing dramatic UK price increases as we are exposed to a developing food crisis in which global prices have risen by over 80% between 2005 and 2008.[36]

Figure 5 overleaf shows that the prices of the items bought by an average UK family (the darker line) have increased by over 30% in the decade to 2008, while the price of their food (the lighter line) has increased by 20%. However, over 2007 this food price inflation was much faster, with the prices of staples like bread and milk rising by about 15% in just that one year. Overall UK food inflation for 2007 was 6%, while inflation for all items was 4%. Although Government figures for 2008 were not available in time for inclusion in this book, the British Retail Consortium has announced that food prices increased by 10% in the year to September 2008.[37]

"The problem is not that the road to Hell is paved with good intentions – it's that the road to Hell is paved."
– Guy McPherson

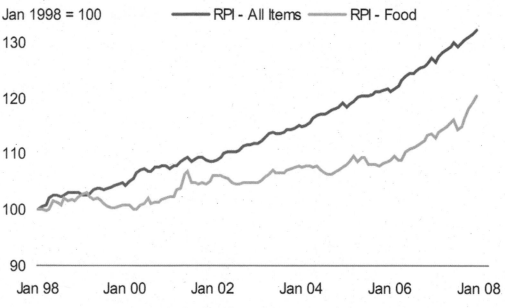

Figure 5: UK food price inflation (RPI = Retail price index).[38]

In a 2008 article, Dr David Barling, Professor Tim Lang and Rosalind Sharpe outline seven fundamental pressures coming to bear on the existing global systems of food supply:

"Land (there's not much more of it); demography (lots more people); fossil-fuel energy (it's near to its limits); dietary change (richer diets leave deeper ecological footprints); climate change (it's set radically to change production and impact on natural resources); water (stress is set to affect billions of people); and urbanisation (more people now live in towns than in food-growing rural areas)." [39]

With such fundamental challenges facing food supply globally, it is pertinent to note that the UK currently produces just under two-thirds as much food as it consumes. Within that, we do produce more cereals than we directly consume, but also currently produce only 5% of our fruit consumption.[40]

Relying on imports for more than a third of our food supply may seem as though it must be a modern trend, but in fact we have not been close to food self-sufficiency since the mid-18th century. Our self-sufficiency ratio increased from a low of around 35% in the 1930s to a peak in the 1980s, but since then has been declining again, as can be seen in Figure 6. It is now among the lowest in the EU.[41]

Of the fundamental global trends just outlined, all are relevant here in the UK, but let

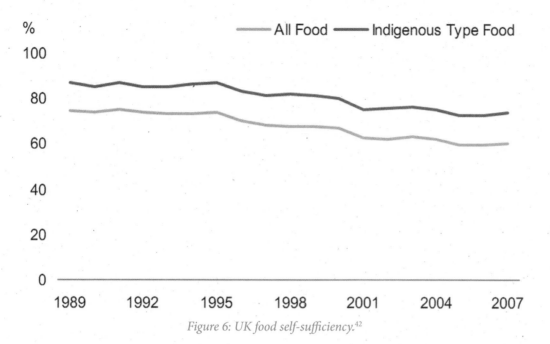

Figure 6: UK food self-sufficiency.[42]

us consider here those which are not touched on elsewhere in this book.

Diet

The UK's diet has been through what is termed the 'nutrition transition' – the change that is seen around the world as people become richer, away from plant-based diets and towards more fat, dairy and meat. We are now eating 50% more meat than we did in the 1960s and considerably more fat and refined carbohydrates. Such a diet not only leaves a much deeper ecological footprint, but also makes us fatter and more prone to heart disease, cancer, diabetes and degenerative diseases.[43]

An urban market garden, part of the Growing Communities initiative, based in Hackney, London

According to the Cabinet Office, an estimated 70,000 premature deaths could be avoided each year if UK diets improved. Obesity and overweight alone currently cost the UK

"Diet is now the single biggest factor in causing premature death worldwide. Even in Sub-Saharan Africa five percent of the population are obese."
– Tim Lang, 'Gardening in an era of food insecurity', September 6th 2008

economy £10 billion a year, and on current trends this is forecast to reach £50 billion by 2050, with a quarter of UK adults being obese by 2010. As many have highlighted, we are eating so much it makes us sick while others around the world suffer malnutrition and starvation.[44]

It must also be mentioned that UK households waste around a third of the food we purchase – 6.7m tonnes (£10.2bn worth) each year. This means that the average household is spending £420 on food it throws out. Local councils then spend another £1bn collecting this and taking it to landfill, with food waste generating greenhouse gas emissions equivalent to 20% of that produced by UK car use.[45]

Water

With regard to water supply, we all know that the UK is a wet country, but it is also a densely populated one. As the Environment Agency highlights:

> "Rainfall is unevenly distributed, both geographically and seasonally. We are not drier than most Mediterranean countries, but we do have much less water per person. Parts of the south east of England have less water per person than Sudan or Syria."[46]

They categorise the whole of south-east England as being under "serious water stress",

and this region is also where the future effects of climate change will be most marked, with the highest predicted rises in temperature and reductions in summer rainfall in the UK.[47]

Nonetheless, the UK as a whole remains relatively 'water-rich', largely because irrigation is generally unnecessary here, while 70% of drinkable fresh water use globally is for agriculture. In contrast, looking globally, while 92% of humanity had a relative sufficiency of drinkable water in 2008, by 2025 this is projected to reduce to just 62%.[48]

Land use

According to the Environment Agency around 72% of the area of England and Wales (about 10,850,000 hectares) is currently used for agriculture. However, due to our high and rising population density, housing and urbanisation are continuing to expand across more of our land, with 31% of new residential land in England in 2004 having previously been farmland. Erosion also removes some 2.2 million tonnes of arable topsoil annually – 44% of the UK's arable soils are suffering from erosion, with 36% at moderate to serious risk.[49]

These trends are increasing the pressure on land availability, leading to choices not only between food for people and food for animals, but also forestry, biomass/biofuels, biodiversity preservation and simply retaining a pleasant countryside.

Image from www.imageafter.com

It is also significant to note that 25-50% of the increase in atmospheric CO_2 from 1850-1990 was caused by land-use change and losses of carbon from soils. At a European Commission conference in June 2008, Professor Rattan Lal of Ohio University presented findings that with changes to agriculture and land use, terrestrial ecosystems could naturally reabsorb sufficient CO_2 to reduce current atmospheric concentration by around 50ppm (the current concentration is roughly 385ppm).[50]

Impacts on farming and farmers

According to DEFRA, over half a million people are currently employed in farming in the UK, but this does not tell the whole story. The Soil Association reminds us that in the early 1900s 40% of the UK population was involved in farming, and that over the past 60 years the number of farmers has declined by over 200,000 (equivalent to 10-12 farmers leaving the industry every day), with the number of farm labourers declining from around one million to under 200,000 (equivalent to more than 35 leaving the land every day).[51]

This has in large part been due to the industrialisation of agriculture, with the energy of human labourers having been replaced by the fossil-fuel energy powering farm machinery and the production of fertilisers and pesticides. This led to more than a tripling in world grain harvests, but also to a decline in knowledge and understanding of farming among the general population. Such industrialised agriculture may require 50 times as much energy as traditional agriculture, but with a barrel of oil containing the energy equivalent of almost 25,000 hours of human labour (and costing considerably less) this presented no obstacle. It has been estimated that 95% of the UK's food is now oil-dependent.[52]

This kind of farming also favours large-scale farms which can afford expensive machinery and are able to make the most efficient use of it. This trend, in combination with the rising cost of farm inputs (and, until recently, the falling price of food) has led to many smaller farms being pushed out of business, and a situation in which the average UK farmer today is approaching retirement age, since few youngsters have seen farming as a promising career.[53]

Government position

The UK Government is showing some belated signs of starting to wake up to the fundamental challenges facing food in 2008,

"Modern agriculture is the use of land to convert petroleum into food."
– Albert Bartlett (1978), 'Forgotten fundamentals of the energy crisis', American Journal of Physics

and the possibility that leaving everything to the markets may not be adequate.

In July 2008 DEFRA published a discussion paper *Ensuring the UK's Food Security in a Changing World* and opened a consultation on the topic, and in 2006, to their credit, they commissioned Dr Helen Peck of Cranfield University to produce *Resilience in the Food Chain*, to begin examining how well equipped our food chain is to cope with "disruption to the supply of money, food, water, energy, fuel, communications or transport, as well as terrorism".[54]

Meanwhile, the Rural Climate Change Forum has been working to raise awareness of climate-change impacts among farmers and landowners, while DEFRA's Farming Futures initiative provides them with practical information and advice. This is essential not only because climate change will affect farming both directly (57% of the UK's best farmland lies below sea-level)

and indirectly (for example through changes in pests and diseases), but also because the food chain currently contributes around 19% of UK greenhouse gas emissions, making food consumption the largest contribution to climate change from the average household.[55]

Perhaps even more importantly, the Scottish Land Reform Act 2003 has given communities first right of purchase when the feudal estates on which they are sited come on the market. This has brought areas of land back under community ownership, which has tended to encourage local employment and food production as well as more sustainable practices. The Community Land Trust model is spreading this movement to England.[56]

Cultural story change

Food and water are so fundamental to human life that our cultural stories in this area are

Food	Land per kg (m²)	Calories per kg	Land per cap p.a. (m²)
Beef	20.9	2800	8173
Pork	8.9	3760	2592
Eggs	3.5	1600	2395
Milk	1.2	640	2053
Fruit	0.5	400	1369
Vegetables	0.3	250	1314
Potatoes	0.2	800	274

Figure 7: Land area needed for different kinds of food production.

especially powerful, deeply-held and central to the ways we live our lives.

Working with local landowners is a key Transition strategy

This is one reason why voluntary change in national diets tends to happen only slowly, but also why the increase in UK consumption of fast food and highly processed and packaged food is so widely derided. The Slow Food movement is (ironically perhaps) growing rapidly, with its emphasis on supporting and enjoying good, clean, fair food.[57]

However, there are perhaps even more compelling reasons for considering a shift in our food habits. Rajendra Pachauri, Chairman of the Nobel Peace Prize-winning Intergovernmental Panel on Climate Change (IPCC), commented in 2008 that:

"The UN Food and Agriculture Organization (FAO) has estimated that direct emissions from meat production account for about 18% of the world's total greenhouse gas emissions. So I want to highlight the fact that among options for mitigating climate change, changing diets is something one should consider." [58]

And as the table opposite shows, meat production, especially beef, also has a huge impact on land availability. It takes 8kg of grain to produce 1kg of beef, and a third of global arable land is currently devoted to producing animal feed. So our diet is not only making us less healthy, it is also negatively impacting our climate and taking land away from other critical uses.[59]

With regard to this issue of land availability, Simon Fairlie recently undertook a 'back of an A4 envelope' analysis of the potential options for UK land use, with a particular focus on the debate between industrialised and organic agriculture. He concluded that Britain has the land to achieve 100% self-sufficiency in food with organic agriculture (as well as making room for our other land demands), but only if we reduce the amount of meat in our diets to around half what we consume today. The perception of heavy meat-eaters could be set to change in much the same way that the perception of 4x4 drivers has done.[60]

Agricultural changes

We saw above that many small farms have been forced out of business due to being economically uncompetitive. This is widely known, but what is less widely understood is that these small farms are actually the most productive per hectare.[61]

"The myth is there isn't enough food to feed people. There is plenty but it is maldistributed." – Tim Lang, 'Food Miles professor calls for shift in food culture at Real Food debate', *Farmers Weekly Interactive*, April 25th 2008

"Food is more than just fuel for the body: it is a source of spiritual, social, cultural and physical nourishment." – Satish Kumar, 'Focus on Food', *Resurgence*, Nov/Dec 2008

David Heath, whose father George ran an extensive market garden in Totnes until 1981, points out the extent of the former garden, now dedicated to car parking

The reason small farms lose out in our modern economy is that, while they can produce substantially more food per hectare, the big farms can produce more of a given monoculture crop per hectare, which suits the large-scale centralised buyers.

In terms of grain alone a big energy-intensive monoculture farm can produce more, but a smaller and more diverse farm can grow several crops simultaneously on the same piece of land, utilising varying root depths, plant heights or nutrient requirements, as well as being able to use perennials, integrate plants with livestock and possibly use draught animals. This all requires more skilled labour and careful management but can produce considerably more per hectare, whether measured in tonnes, calories or income. Indeed, if they can access buyers, small farms can produce huge cash yields per hectare by focusing on highly priced produce.[62]

This accounts for the famous fact that the most productive parts of most farms are actually the farmers' gardens, where far more human care and labour is invested in the diverse yield. Clearly such human-intensive production is also vastly more energy-efficient. In this sense, it is gardeners, not farmers, who are the most appropriately skilled food growers post-peak oil, especially since gardening has become a key cultural repository of skills in the absence of a farming culture. Richard Heinberg has written of America needing 50 million farmers, but it would be a great start to make the UK a nation of 50 million gardeners.

Getting many more food consumers involved in food gardening would not only improve food security, but also encourage important skills, improve health and reduce some 'food miles' to 'food yards'.[63]

Walnuts – a much under-utilised potential food source

Communities and individuals can also support and benefit from local farming through participating in the wide variety of Community Supported Agriculture (CSA) schemes – investing in a farm in some way in return for a share of future harvests – and by purchasing food directly from local farmers at farmers' markets, thus ensuring that farmers receive all of the money spent on that food, rather than the small percentage they receive on supermarket-bought food. These approaches also encourage stronger community ties and the consumption of locally-grown, seasonal produce.[64]

The global picture

All of these changes to our thinking around food at the local and national level would, in our globalised world, help considerably with the challenges facing food globally.

It is not so much that food supply is a problem only for poorer countries, but rather that our food decisions in the rich world are *causing* these problems in the poorer countries. By relying on free markets to manage distribution and on using our wealth to buy their food we are creating high prices, and thus shortages.

Another critical issue here is our role in the global water situation. Meat and dairy-based diets have a huge 'water footprint', with 1kg of grain-fed beef requiring 15 cubic metres of water compared with 1kg of cereals needing only between 0.4 and 3 cubic metres. This is known as 'embedded' or 'virtual' water and we are effectively importing this water from other countries which are under much greater water stress than our own. According to the WWF, the UK effectively imports 62% of its water footprint, and in his book *When The Rivers Run Dry*, Fred Pearce calculates that the equivalent of 20 Nile rivers move from developing to developed countries each year.[65]

A poster from a Transition Town Lewes event. Helping people to grow food again is seen as being a high priority for Transition initiatives

"The world has a surplus of food, but people still go hungry. They go hungry because they cannot afford to buy it. They cannot afford to buy it because the sources of wealth and the means of production have been captured, and in some cases monopolised, by landowners and corporations. The purpose of the biotech (GM) industry is to capture and monopolise the sources of wealth and the means of production."
– George Monbiot, 'Seeds of Distraction', *The Guardian*, March 9th, 2004

Genetically Modified (GM) foods

Some suggest that GM foods are the solution to world hunger, but it is hard to see how they will resolve the fundamental problem of distribution and inequity. As Professor Lang puts it:

"There's a probability of GM offering a solution to what it cannot resolve. It cannot resolve water and energy shortages. It cannot resolve health. It cannot resolve the impacts of climate change. It might tweak a bit here and there. GM is a side issue. It doesn't address the real fundamentals. Most uses of GM are to preserve the agrochemical market." [66]

Many questions in agriculture, including those around GM, have recently been the subject of the UN-backed International Assessment of Agricultural Science and Technology for Development (IAASTD) report [67] – the result of four years' work by over 400 scientists and the biggest study of its kind ever undertaken. After its conclusion the Director of the Assessment, Professor Robert Watson (also DEFRA's Chief Scientist), was asked if GM could solve world hunger, and replied:

"The simple answer is no." [68]

Indeed, there is some evidence that GM crops may even reduce yields, as found by two separate studies at the University of Kansas and the University of Nebraska. The IAASTD report also found that industrialised agriculture is "too narrowly focused" and called for more support for small-scale farms, local farmers and traditional knowledge, finding that, as in so many other areas: [69]

"Business as usual is no longer an option." [70]

The Transition Vision – looking back from 2027

From the Second World War through to the end of the 2000s the simple aim of UK Government food policy was to ensure that enough food was available, affordable and accessible. At first glance these aims appeared reasonable enough, but by 2008 it was becoming clear that there were other critical concerns – our food supply also needs to be healthy, resilient, low-carbon, low-water, low-energy, low-land-use, supportive of ecosystems and sound employment, technologically appropriate and socially just!

The 'Edible Garden Crawl' tour in Totnes, looking at a range of food gardens

In light of growing costs and public dissatisfaction, the Government recognised the need to take back some of the responsibility for the management of the food supply chain, which it had essentially been leaving to the supermarkets. In 2010, the Government outlined its new Vision For UK Food, to sit alongside the EU's new Common Sustainable Food Policy. To the surprise of many, this was broadly welcomed by the major supermarkets, which were becoming increasingly uncomfortable at bearing the brunt of discontent from consumers and farmers alike.

A central pillar of the Vision was the response to climate change, and a recognition of the need to increase levels of organic matter in UK soils. DEFRA's Farming Futures initiative started advocating the Permaculture Principles in their advice to farmers for the first time, and encouraged no-till farming systems, which leave crop residues on the fields and avoid the use of herbicides. The Vision also mandated a faster expansion of forests and woodland in the UK, recognising their many benefits in producing fuel, food, industrial products, carbon sequestration, ecosystem habitats and happier people.[71]

As part of the Government's new Vision, import controls were introduced on foods that could be produced in the UK and local authorities were encouraged to shift their procurement policies towards local and seasonal food. It also supported the Community Land Trust model, and pushed through a big public education drive regarding the wide-ranging impacts of our choices in diet, water use and food purchasing, which served to reinforce the growing trends towards farmers' markets and community-supported agriculture.

By 2013 the new Vision was having a clear effect on consumer choices, with demand for meat beginning to reduce, leading to a reduction in pressures on land use. Organic sales were also increasing rapidly, but this was primarily due to the fact that by this time most organic foods were cheaper for consumers than the industrialised equivalent, due to the rising cost of energy. Meanwhile, the Slow Food movement continued its inexorable spread, with over 1,000 local 'convivia' encouraging and supporting good, clean, fair food across the country.

Many Transition initiatives organise seed exchanges

Urban food production is now an established part of the picture, with allotments and gardens greatly increasing our self-sufficiency in fruits, nuts and vegetables. Garden-sharing has also become a common arrangement, with those with the time and skills cultivating underused gardens in exchange for a share of the produce. Rainwater harvesting for watering plants is also the norm, with increasing numbers also using rainwater in their toilets and washing machines.

The Royal Parks have become emblems of our new attitude to food, with the majority of food sold in park cafés grown on site in attractive displays, and local schools also among the beneficiaries. Meanwhile, the majority of urban trees planted since 2010 have been food trees, and pictures of London from the air are now considerably greener, with food growing on rooftops, in window boxes and even up walls!

In 2027 the UK's food self-sufficiency ratio is over 96%. The biggest farms have largely broken up due to their inefficiency, and have been replaced by a diverse range of small-scale farms, leading to a substantial increase in productivity and increased employment. These farms have often formed farmers' co-operatives, and have again become integral hubs of their local communities, with most of their produce consumed locally – not only food, but also biofuels and biomass, local building materials, medicinal plants, plants for fabrics and even heat and power. Farms and food have again become a key part of most people's lives.

Chapter 9

Electricity and energy

Present position and trends

Around 90% of the UK's energy needs are met by oil, gas and coal, and our ability to produce or import these fossil fuels is likely to become significantly more constrained over the period to 2027 (pp.156–161). Renewable energy use is now over five times the level it was at in 1990 and growing fast, but in 2007 contributed only 3.3% of our energy demand.

Looking just at electricity, 78% is generated from fossil fuels, with oil contributing 1%, natural gas 43% and coal 34% of our total supply. Nuclear electricity has been declining for the last decade, and will continue to do so over the next twenty years, but still contributed 15% in 2007. Renewables continue to grow rapidly but contributed only 5%. Electricity generation currently accounts for about one third of the UK's total carbon emissions.[72]

In the UK both electricity demand and supply had been increasing for decades, but with demand over the past couple of years curbed by mild winter weather and high electricity prices, there has been a modest drop in both demand and electricity available for supply between 2005 and 2007, although the Government expects demand to start rising again in future years.

Importantly, 22.5 gigawatts (GW) of existing coal and nuclear power stations (around 30% of the current total generation capacity) are scheduled to close by 2020, mostly due to either reaching the end of their planned lifetimes or being too polluting to continue under new EU environmental legislation.[73]

For those who like graphical representations, the energy flow chart in Figure 8 (from the Department for Business Enterprise and Regulatory Reform) provides a high-level overview of energy in the UK in 2007, illustrating the flow of primary fuels from the point at which they become available from production or imports (on the left) to their eventual final uses (on the right). Flows at the bottom represent exports, conversion losses, energy industry use and non-energy use. The blocks with cross-hatching represent transformation (power stations and refineries).

The number of UK households in fuel poverty – needing to spend more than 10% of income on fuel to keep the home at an adequate temperature – had fallen from around 6.5 million in 1996 to around 2 million in 2004. Yet the recent increase in energy prices has turned this trend around. The latest available Government figures show a rise to 2.5 million fuel-poor households in 2005 and likely over 3.5 million in 2006. The consumer group Energywatch claim that by January 2008 there were 4.4 million fuel-poor households,

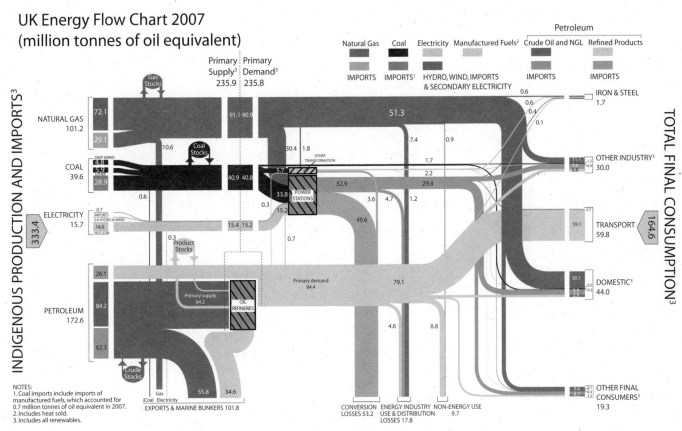

Figure 8: UK Energy Flows 2007.[74]

Source: chart produced by the Department for Business Enterprise and Regulatory Reform.

and indeed it would be surprising if this upward trend had not continued, with domestic energy prices increasing by 26% between 2005 and 2007. Petrol prices increased by 8.5% over the same period.[75]

It is clear that our existing energy strategy is under strain, and that many are already suffering the consequences of this. Continuing down this path is deeply undesirable, and as Part Five's exploration of the UK impacts of peak oil and climate change shows, would only lead to a massive economic burden on the country and likely also to energy shortages of many kinds. It is apparent that we need a change of direction.

Our Government is starting to show an awareness of these issues (see pp.156–158), but is not responding nearly urgently enough.

One hopeful sign is the creation of a new Department of Energy and Climate Change in October 2008, which has the opportunity to do the kind of joined-up thinking which is necessary to create the kind of future we all want to live in.[76]

Cultural story change

As discussed above, around 30% of the UK's current electricity-generation capacity is scheduled to close by 2020. The current discussion revolves around what can replace this lost capacity and provide for the anticipated growth in demand, yet we are running out of desirable options.

The obvious option is renewables, and we do clearly need to increase renewable energy capacity as quickly as possible. The Government has set a target of generating 10% of electricity from renewables by 2010, but looks likely to miss this target. Europe has set the commendably ambitious target of 20% of its electricity, heating and transportation running from renewable sources by 2020, and the UK's commitment towards this Europe-wide goal is 15% by 2020. However, the UK Renewables Advisory Board (in common with other studies) estimate that with current policies the proportion of energy from renewables will only reach 6% by 2020, and that even with radical policy changes and great effort they can only foresee our reaching 14% by 2020. Serious social and economic change is required to go further. As Alan Moore, co-Chairman of the Renewables Advisory Board, put it:

"If the 15% target is to be approached we need to establish a different energy world." [77]

We do indeed, and not only to meet our renewables target. If renewables cannot replace the power stations scheduled to close, then the obvious question is what can? We have examined in depth the environmental and supply challenges facing fossil fuels (the talk of 'clean coal' is more marketing than reality, which is why the EU are shutting down dirty coal power stations), but what of nuclear? [77a]

As it turns out, nuclear electricity generation faces many of the same problems, as well as its own unique difficulties. The depletion of high-quality uranium ore means that the Energy Return On Energy Invested (see p.121) of nuclear electricity is worsening, to the point where if we take into account the full life cycle of nuclear electricity generation (including building and decommissioning the power stations, mining, transporting and milling the fuel and managing the waste), it becomes an open question whether future nuclear projects are likely to generate more energy than they use. It may be that nuclear is actually an *energy sink*, not an energy source (in other words, actually worsening our peak oil/climate change challenge), as well as committing us to millennia of high-tech nuclear waste management, exposing us to the possibility of nuclear accident and increasing the risk of the proliferation of nuclear weapons. It is also worth remembering that nuclear electricity is far from carbon-free, since most aspects of the nuclear life cycle are powered by fossil fuels.[78]

So if neither renewables nor nuclear can fill the 'looming gap' in our ability to meet energy demand, and we don't want the consequences of more fossil fuel power stations, what can we do? The Government's current answer is essentially a simple one – pray to the market:

> *"We believe that a market-based approach is the best way to manage these uncertainties, providing the flexibility to be responsive to developments we cannot yet know."* [79]

Yet there is another option that is rather more likely to succeed as a response to a shortfall in energy supply – simply reducing our energy demand.

The Transition Town Lewes office is also home to the Ouse Valley Energy Services Company (OVESCO)

When we talk of demand and supply as two separate and unrelated factors we are led back towards the need for more energy to meet demand in a world of depleting energy resources. But in reality how much energy we need is governed in part by how much energy we have, and how we choose to use it. The more feedback there is between suppliers of energy and users of energy, the more likely we are to use it well.

We need a new relationship with our National Grid that makes this connection apparent to everyone, helping us to understand the challenges we face and consider them in our everyday thinking – 'shall I use the washing machine at the point of peak demand, or do it later when the system is under less strain?'

Rebecca Willis has laid out this possibility in her brilliant report *Grid 2.0*:

> *"In Grid 2.0, much more power will be generated at community and household level through renewable and low-carbon technologies like solar and wind power, small-scale combined heat-and-power, heat pumps and biomass boilers. There will still be large-scale power generation, especially for industrial use. But the National Grid will transform from a one-way provider of power to consumers, to a two-way web linking distributed sources of energy supply and demand. . . . the National Grid will become an enabler rather than an automatic provider of power, linking microgrids and allowing distributed generators to trade with each other, in order to even out supply and demand."* [80]

If we start to see the Grid as an enabler of relationships between suppliers and users, rather than as an all-powerful provider, we are likely to see demand adapting to supply, whereas at present we see only supply straining to adapt to demand.

> *"There is a widespread assumption that a consumer-capitalist society, based on the determination to increase production, sales, trade investment, 'living standards' and the GDP as fast as possible and indefinitely, can be run on renewable energy . . . But if this assumption is wrong, we are in for catastrophic problems in the very near future and we should be exploring radical social alternatives urgently."*
> – Ted Trainer (2007), *Renewable Energy Cannot Sustain a Consumer Society*, Springer Verlag

"The most powerful energy resource we have available to us is the creative intelligence of the people"
– David Fleming, inventor of TEQs

The other key framework in changing our cultural stories around energy and encouraging demand reduction is TEQs (Tradable Energy Quotas). This is the system of energy rationing for individuals and organisations currently being examined by the UK Government, which could provide a way to guarantee that we achieve our emissions reduction targets, as well as empowering communities to address their energy challenges.[81]

Rationing has acquired a bad name due to its association with shortage, yet it is a response to shortage, not the cause of it. The word 'rationing' really contains two related but distinct meanings – guaranteed minimum shares for all on the one hand, limits to what people are allowed to consume on the other. Many of us resent the second, but in times of shortage we cry out for the first. Fortunately, TEQs are rationing in the first sense only, as they guarantee minimum shares for all, but do allow individuals to exceed their basic entitlement (if they are willing to pay those who do not for the privilege). Introduced in time, they would allow us to sensibly manage a decrease in energy demand, rather than carrying on blindly and panicking when we hit the wall.

Under the TEQs scheme, every adult is given free energy rations, and all organisations (including the Government) must buy rations to cover their energy use. No energy can be used in the economy without the requisite rations, and the total number of rations issued is set by the carbon emissions budget of the country concerned, meaning that the achievement of emissions targets is guaranteed. These rations are tradable within the country, so that the energy-thrifty can choose to sell their surplus rations, earning a profit for themselves and creating additional flexibility in the system.

The beauty of this system is that it not only stimulates awareness of our energy challenges among all energy users, but it also creates a clear indicator of how well our society is doing in the move towards a satisfying low-carbon existence. The national price of TEQs units will be a part of our everyday lives, and it will be in everyone's interest to keep it as low as possible, whether by working and co-operating to reduce energy demand or by creating new low-carbon sources of useful energy.

Many Transition initiatives set up solar buyers' clubs – a good way of getting discounts on solar installations

Frameworks like TEQs and Grid 2.0 can help to create a society in which it is very ordinary to be concerned with improving efficiency, relocalisation, and renewable low-carbon energy generation. The Centre for Alternative Technology's *Zero Carbon Britain* report argues that via the TEQs system it is feasible to reduce

UK energy demand by 50% by 2027 without any loss in quality of life (as conventionally defined) – in the current global context this appears a far more sensible approach to the so-called 'energy gap' than trying to fill it with more nuclear, coal or gas power stations.[82]

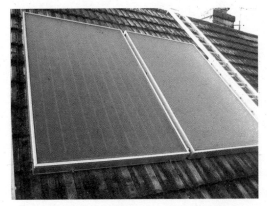

Solar thermal, often the most cost-effective micro-renewable

The Transition Vision – looking back from 2027

As the UK's energy imports became ever greater and more expensive from the late 2000s, and more households suffered the consequences, the Government was forced to look again at the assumptions underlying its energy policy. Also, with the TEQs system implemented in order to achieve our strict climate commitments under the '2010 Accord' (see p.33), the general public became a lot more aware of how much the current energy system was truly costing them, and the pressure on Government to find better options increased. The EU's target of 15% of UK energy from renewables by 2020 was also increasingly slipping away.

As such, in 2011, emboldened by the surge in interest in the area, the Government announced that it was "getting serious about microgen", and that from then on it would compare the cost of proposed new generating capacity (and its fuel) with the cost of *saving* the same amount of energy elsewhere – the 'negawatt' principle – and factor in the value of the diversity and resilience offered by small-scale systems. This was widely seen as an acceptance that there would be no more new coal-fuelled power stations, and celebrated as a victory by the massive direct-action movement that had been blocking proposed developments.

A photovoltaic bicycle shelter in Machynlleth, Wales

The Government also prioritised National Grid connections for small-scale energy producers and began rewarding the operators of the Grid for connecting more distributed generation and reducing transmission losses, rather than simply for increasing the throughput of electricity.

Retrofitting buildings using local materials in an area of interest to Transition groups. Here a wall is being insulated using a mixture of hemp and lime.

Meanwhile, the big electricity suppliers gradually moved towards time-of-day pricing, reflecting the challenges they faced in providing power at times of peak demand. In response to this, 'dynamic demand' technologies instantly became popular with consumers, automatically switching off non-time-critical appliances like fridges and water heaters during times of high prices (and peak use). Cheaper tariffs for industrial customers who would accept a supply that could be interrupted during unexpected peaks in demand were also introduced.

With the rising price of energy, and with everyone's use of TEQs units also tied to their energy use (and to the carbon emissions associated with that energy), the UK public became very keen on energy-saving, and ingenious solutions for reducing energy usage became a popular topic in conversations and newspaper articles. The Government also supported the creation of 'energy mortgages', in which the set-up costs of small-scale energy generation are paid back through the mortgage.

Almost all homes and businesses had 'smart meters' by 2011 too, providing a highly visible live readout of both the up-to-the-minute electricity cost and the energy use and energy generation currently taking place in the building. As these were installed, people were often surprised by which of their activities used the most power, and pleased to find how easy it was to make initial big savings through small changes in habits. The sight of unoccupied office blocks lit through the night is a fading memory.

The UK population peak in 2013 and the beginning of its gentle reduction (see p.48) was also an essential part of keeping total energy demand down. Increases in energy efficiency had previously been largely overwhelmed by increasing consumption, but now with both a shrinking energy footprint per person and a shrinking population, living within our energy means became achievable.

It has also become increasingly apparent that communities – from estates to neighbourhoods and villages – represent the best scale for collective solutions to heat and power generation, and by 2023, the majority of energy assets were at least part-owned by the local community, often through co-operative renewable energy societies based on the Totnes model. In 2025 the Government also introduced a requirement for any new large-scale generation to be partly community-owned. Local insulation clubs and TEQs-pooling groups are also common and every local authority has a detailed energy strategy. Ironically, although Grid operators are now incentivised to connect small-scale producers, many producers see no urgent need to hook up and sell electricity into the national system as they are using it more locally.

By dramatically reducing the amount of electricity required, we actually achieved over 30% of UK electricity generated from renewables by 2020, and with the decline of North Sea oil and gas (by 2027 down to one-sixth of 2007 production levels), and the extreme difficulty in sourcing foreign supplies, by 2027 over 70% of our electricity is being produced renewably, simply out of necessity. While other European countries like Denmark have already achieved virtually 100% renewable electricity, we are playing catch-up as our last fossil-fuelled power stations move towards the end of their lifetimes.

Chapter 10

Travel and transport

Present position and trends

Around 95% of transport around the world is fuelled by oil, and over two-thirds of the UK's oil consumption is now fuelling transportation. The total energy consumption of UK transport has almost doubled since 1970 and currently represents around a third of the UK's primary energy use.[83]

Total road traffic has increased by 84% between 1980 and 2006, from 277 to 511 billion vehicle kilometres. In 2006 bus/coach travel accounted for 5.4 billion km, commercial light van traffic covered 64 billion km, and Large Goods Vehicles (lorries over 3.5 tonnes) 29 billion km. Motorcycle use has been increasing in recent years, as has cycling, which accounted for 4.6 billion km.

Nonetheless, a full 80% of UK road traffic is cars, which accounted for 686 billion 'passenger kilometres' in 2006. Average car occupancy has remained roughly stable at around 1.6 persons per vehicle since 1995, with most journeys undertaken by cars occupied only by the driver (over 30% of UK households now have two or more cars).

UK traffic.

Source: istockphoto.

By comparison, in 2005/06, rail passengers travelled 43 billion passenger kilometres, an increase of 43% since 1980. Passenger kilometres travelled by air have more than trebled in this time, reaching 191 million in 2004, with international air traffic growing at a slightly faster rate than domestic. In 2006 holidays accounted for nearly two-thirds of UK residents' trips abroad (45 million holiday departures), with 81% of all our trips abroad taken by air, despite that almost three-quarters of these were to European destinations (19% to Spain alone).

Freight transport has increased by 44% since 1980, to 252 billion tonne kilometres, with 66% of this travelling by road, 21% by water, 9% by rail and virtually all of the remainder by pipeline.[84]

These transportation choices have reshaped our landscape and even changed the air we breathe, with roads and railways now the most apparent divisions snaking our countries, and road transport the largest contributor to airborne pollution.

And as Sustrans' *Towards Transport Justice* report comments, the overwhelming dominance of the car and all the infrastructure that surrounds and supports it also has deep social justice implications:

> *"Injustice is experienced by millions of people in the UK who do not have a car, or struggle to afford one, in the 'must-have-car' society which we have created. These injustices include difficulties accessing work and other opportunities; enforced indebtedness; reduced opportunity to lead an active healthy life; and, ironically, proportionately greater exposure than average to pollution, road danger and noise caused by those who do have a car."* [85]

Around a quarter of UK households do not have access to a car, including the majority of the poorest 20%. Those in this income bracket that do have a car spend nearly a quarter of their income on the cost of motoring. Meanwhile, 40% of jobseekers say that lack of transport is a barrier to getting a job, and inaccessibility of work is cited as the most common obstacle to young people gaining employment, while

over a million people annually miss out on medical care because of transport difficulties. Our current transportation system is failing those who do not have access to a car.[86]

Our current transport system discourages cycling, but it could replace driving for many trips.

The last three decades have seen the cost of motoring fall by 10% in real terms, while average household disposable income has more than doubled, yet with rising fuel prices we are already seeing that more and more people will struggle with the cost of car ownership, leading to more of this kind of social division. Those whose lives are currently entirely car-dependent will clearly be the worst affected, and a report for the AA and the UK Petroleum Industry Association found that:[87]

> *"Poorer households . . . that do run cars . . . are particularly hard-hit by fuel price increases. For instance, our estimate is that when fuel price goes up, in order to preserve the volume of fuel they buy, a person living on their own on a state pension and paying for petrol would have to spend twice the proportion of*

household income as a two-adult family or a retired couple not on a state pension." [88]

These kinds of financial pressure are already leading to rising rates of theft of all kinds of fuel. In August 2008 Channel 4 News conducted a survey of police forces across the UK and found that fuel theft from vehicles was up 38% compared with the previous year (400% on Tayside), theft of red diesel used in agriculture was up 48% and thefts of home heating oil up 222%. [89]

Even without considering climate change we can see that our current transportation systems are in need of radical restructuring. Yet, as the UK Climate Impacts Programme highlights, our transport infrastructure is vulnerable to disruption from extreme weather events such as flooding, strong winds and fog disrupting air travel, and rails buckling in extreme heat.

The climate change outlook means that we can only expect such events to occur more frequently, yet we are still contributing to this growing problem, with our Government supporting expansion of both airports and our road networks. In 2004 road transport accounted for over 20% of UK CO_2 emissions, and a Department for Transport-commissioned report forecasts road transport emissions increasing by 25% by 2030. [90]

Cultural story change

Our cultural stories around transport are very much based on the right to freedom, but as we have seen, the freedom provided by our current infrastructure is only available to those who can afford it, and this group is shrinking. We could instead strive to create an 'equal opportunities' transport system, which creates freedom and health for everyone.

We must also recognise that this right to freedom carries with it responsibilities. As well as the obvious contribution to climate change and oil consumption, our perceived right to rapidly travel anywhere in the world at no notice must be weighed against adverse effects, such as increased air pollution levels and the number of deaths and injuries on our roads (both considered in the Health chapter, p.79 onwards).

The key insight here, though, is that while what we actually desire is *accessibility* (the ability to access the goods and services people travel for), UK transport policy to date has focused on increasing *mobility* (the ability to travel further and faster).

This focus is strange, since increased journey distances are not in themselves desirable, and they bring with them increased motor traffic and fuel use. Increased speeds obviously increase the risk of accidental injuries and fatalities, but also significantly impact fuel usage – the energy needed to push an object (such as a vehicle) through the air increases as the cube of velocity (x^3), so if we double velocity the amount of fuel-energy we use increases *eightfold*. Many other factors also contribute to the fuel-efficiency of travel, but increased speed is intrinsically more energy-intensive, whatever form of transport we are considering. [91]

Consequently, a more sensible aim would be to use the *minimum* mobility to achieve *maximum* accessibility. Interestingly, research

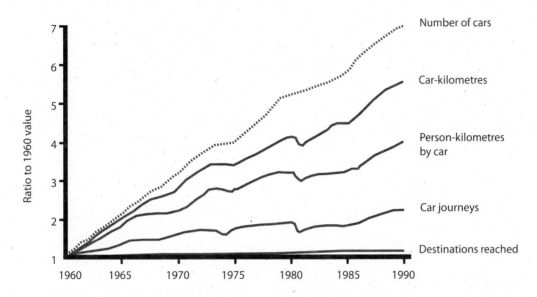

Figure 9: West Germany, 1960-1990, showing typical travel trends.[93]

from around the world shows that as mobility increases, the number of trips does not – on average, people make about 1,000 trips each per year (regardless of their wealth, location or local infrastructure), with over 80% of those trips beginning or ending at home. In other words, it appears that increases in mobility do not increase our access to the things we want, they just mean we travel further and faster to get to them. We could be free to enjoy our lives just as much by accessing our needs and desires more locally, while also being free of the many adverse consequences of our current transport infrastructure.[92]

This means that the kind of fun, healthy, fossil-fuel-free travel celebrated at events like the London Freewheel, Ciclovías and the World Naked Bike Ride could indeed be a big part of the future, and could provide us all with both the freedom to live our lives to the full and a more pleasant environment in which to do so.[94]

The Transition Vision – looking back from 2027

While oil prices began fluctuating wildly in the late 2000s, the trend was clearly upward, and steadily rising forecourt fuel prices reflected that. With this ever-increasing financial cost of running a car, more and more people

"Transportation is not an end — it is a means to having a better life, a more enjoyable life — the real goal is not to improve transportation but to improve the quality of life."
– Enrique Peñalosa, former mayor of Bogotá

"A journey is best measured in friends, rather than miles."
– Tim Cahill

were forced to make do with the inadequate alternatives available at that time.

The first major business casualties of rising fuel prices, though, were the budget airlines, who rapidly lost customers as they were forced to raise fares to cover their growing costs. In 2010, the Government announced that they were 'shelving' their airport expansion plans, partly in response to airline closures and partly in line with their tough new international climate commitments. There was an unspoken understanding that the growing environmental direct action movement had also played a strong role, and there has never since been any talk of 'unshelving' airport expansion.

With the proportion of non-drivers growing substantially and increasing alarm over climate change, there was a growing sense of 'us and them' in the UK, and being a lone driver in an inefficient car increasingly came to be perceived as irresponsible and even unacceptable. It is amazing now to look back and think that as recently as 2007, 25 SUVs were sold here for every one electric/petrol hybrid car.[95]

In light of these shifting attitudes the Government was forced to re-evaluate its 'voters = motorists' mindset and recognise that even those who are sometimes motorists are also pedestrians and users of public transport, as well as using public spaces and breathing public air. The introduction of the national TEQs system in 2010 also led to much stronger demands from both individuals and businesses for low-carbon, low-energy transport solutions.[96]

Considering these factors along with the challenging economic outlook, it became clear that the Government could neither justify nor afford spending £1 billion a year on building new roads, and in 2011 a new transport framework was announced. This included a reduction of the speed limit to 60mph from 2012 (to maximise engine efficiency), increasingly tight fuel-efficiency standards and investment in the capacity of our canal system to enable it to carry more freight. Ambitious aims to complete the electrification of the rail system and to electrify 95% of road traffic by 2027 were also set.[97]

While road maintenance remained a priority, the majority of road expansion projects were indefinitely 'postponed', with a wide range of measures instead introduced to shift people to oil-free transport (walking and cycling) where possible, and to vastly improve the alternatives to car travel. The capacity of the rail services was singled out for improvement, but the most visible and controversial step was to introduce a dedicated bus and coach lane on the biggest motorways, which almost single-handedly made travelling intercity by coach faster than going by car.

This met with outraged opposition from motoring groups, and was nearly overturned, but once people had become accustomed to the speed, convenience and low price of the coach services it became a popular option, not least due to the ability to work or play online while travelling via their wireless broadband connections.

Over the next decade the Government invested in thousands of small-scale improvements to the transport system, from increasing tram services and reopening local rail links to extending the National Cycle Network and expanding pedestrianised areas in many city centres. Many local councils also allocated more road space to buses, cyclists and pedestrians.

A free bicycle repair workshop, organised by Transition Bristol

By 2020, air pollution, road noise and traffic injuries and deaths were all greatly reduced, which only encouraged greater interest in walking and cycling. This led to a 'virtuous cycle' as the greater demand led to better infrastructure (bike-racks, lockers, workplace shower facilities), which in turn led to yet more people reclaiming the streets. Similarly, using public transport late at night or in remote areas become much more appealing as usage rose.

The most unforeseen aspect of the transport transition though was the ingenious methods people invented to get themselves around while using less energy. The TEQs scheme effectively meant that anyone who could get by with less would be rewarded twice, with reduced fuel costs and more spare TEQs units to sell, and the collective genius of our communities was nowhere more apparent than in responding to this.

The sheer variety of forms of transport is one obvious manifestation, as is that of 'E-thumbers' tapping information into their phones at the roadside. They are registering their desired hitchhiking destination to the national open source Lift-Hiker system. With GPS (Global Positioning System) technology now integrated into both mobile phones and car navigational aids, any driver who wishes to fill space in their vehicle simply presses a button, and any nearby person who wants a lift in that direction is sent the details of the driver's name, type of car, number-plate and a suggested rendezvous time and location.

The system is now used by the majority of the population, and has led to far higher vehicle-occupation rates and consequently lower energy costs for both individuals and the country as a whole. Many businesses,

"One's destination is never a place, but a new way of seeing things."
– Henry Miller

groups and communities also buy an electric car between them and then use it on a 'pay as you go' basis, often sharing the cost with E-thumbers picked up along the way. Between them, buses and Lift-Hiker provide a reliable and highly-flexible low-energy transport service for the places where rails do not run, as well as stimulating diverse social interaction – we all know couples who met through Lift-Hiking!

Ironically, considering the controversy when the idea was first discussed, when the Government again reduced the speed limit for oilcars to 50 mph in 2018, nobody much noticed, as the so-called 'hypermiling' trend had long since taught drivers that fuel efficiency dropped rapidly at speeds higher than this.

In 2027 urban areas now average one (electric) car for every ten households, and the cleaner air and relative lack of danger has meant that children are once again able to run, play and cycle on the streets around their homes, although they do sometimes elicit angry words when their games stray onto the cycle tracks. Oilcars are still seen relatively frequently, but are on the whole regarded as dirty relics of a bygone age, kept running by rich hobbyists.

85% of intercity travel is now by train or coach (both electric) with some high-speed rail links available, but aviation is confined to a few large aircraft making occasional flights between the remaining airports. The additional cost of the TEQs units for their jet fuel (including the additional weighting for the climate damage caused by high-altitude emissions) proved to be the last straw for mass rapid flight, with the vast majority of overseas travel now taking place by sea.

One of the big trends here is the 'slow travel' movement, with people returning to the ethos of 'the journey being half the pleasure' and immersing themselves in the culture of their destination upon arrival. Airship cruises, travelling by canal boat and sailing on 'pirate galleons' are especially popular options.[98]

Overall, people are simply travelling less far (and less fast) as they go about their days. Zero carbon travel has increased massively and electric-powered transportation provides for virtually all of our remaining needs. As well as online solutions for business, and more jobs which are able to be performed entirely online, a far higher proportion of shopping is done through the Internet, and the (electric) delivery systems are far more co-ordinated and efficient, with regionalised production, load consolidation at distribution centres and freight volumes and weight reduced through good design and technological innovation.

The environments in which people live have also generally become far more pleasant through these changes, and local councils are organising and encouraging far more events and festivals to reduce people's need to 'get away from it all'.

DAILY EX

THE WORLD'S GREATEST NEWSPAPER

www.express.co.uk

PRINCE OPENS NEW TATE

PRINCE CHARLES yesterday opened the new Sunderland Tate Gallery, having travelled to the city on the 'M1 Rocket', the coach system introduced so successfully last year, where one lane of the motorway is now dedicated exclusively to coach travel.

He arrived looking refreshed and rested, where he was greeted by hundreds of well-wishers. The Prince, who last year gave up travel by helicopter and aeroplane, has taken his role of 'Post-Oil Royal' very seriously.

He was delighted with the new Tate, a zero-carbon building which sports a glazed market garden on its roof. Museum curator Jasmine McIntyre conducted the Prince's tour of both the gallery and the garden, which supplies much fresh produce to the surrounding community. "He was thrilled by the depth of creativity the Tate embodies on all levels," she said.

The Prince denied that, ultimately, he had found the garden more interesting than the art, although onlookers said that he had appeared less than impressed with Damien Hirst's new piece, 'Scrotum Pole', which takes up much of the main gallery.

THE SUNDAY TIMES
timesonline.co.uk/travel

Holiday 2018
Staying at Home is the New Going Away

TWO WEEKS IN TUSCANY IS JUST SO 2010!

All the rage now for the discerning holidaygoer is staying at home. In the pursuit of the low-carbon, non-TEQs-busting, perfect two-week break, thousands are now looking no further than their own place. Holiday advisor Gisella Hawkin gave the Sunday Times her eight tips for the perfect stay-at-home holiday.

- Lock away all communication devices, laptops, palmtops, mobiles, Z-phones and chat-hats
- Time your holiday so that it falls at a time where your home plot is brimming with vegetables
- Visit all the local places you have never visited, museums, parks, theatres, restaurants
- Take bike rides
- Take some time to read the pile of books you spent the previous year putting to one side for when you had the time to read them
- Start a list the previous year of all the things you would do if you had the time, and then design your two weeks around them
- Use the TEQs you have saved by not travelling to treat yourself to a visit from an aromatherapist, a masseur, or even a chef for the night!
- Do a painting or study course, you could even get a neighbour round to pose for their portrait!

Gisella's new book, *Home Sweet Home: the ultimate stay-at-home travel guide* is published this week by Smart Future books, **priced £16.99.**

Chapter 11

Health and medicine

Present position and trends

In keeping with the rest of our society, our healthcare system has become ever more reliant on oil and cheap energy, and is currently little prepared for a lower-energy future.

Much of the electronic equipment used in modern medicine – especially large machines like MRI scanners – is highly energy-intensive and the NHS's buildings tend to be inefficient and poorly-maintained, with total energy use in NHS healthcare facilities costing around £400 million annually. A New Economics Foundation report in 2007 estimated that better building design could cut NHS energy costs by a quarter.[99]

Medical products such as gloves, syringes, IV and dialysis tubing, tablets, gels, ointments, antihistamines, antibiotics and antibacterials are not only petrol-based, but also are never reused or recycled, while many products are also double-packaged in order to maintain a sterile environment. Disruptions to the supply of these products (such as occurred during the 1973 oil crisis[100]) are clearly critical issues. In addition, the NHS's drug budget is increasing by 7.5% a year (at which rate it will double in less than ten years), with attendant increases in energy use in production, development, and distribution.[101]

Also, just as our dependence on industrial agriculture has led to a widespread decline in food-growing skills, our healthcare system's reliance on energy-intensive machines and equipment is also leading to a decline in certain medical skills. In orthopaedics, for example, doctors used to largely use touch to diagnose a problem, but nowadays X-rays or fluoroscopes are used, so new doctors don't gain the same experience and skill, meaning that they may become less able to function effectively outside of a high-energy medical environment.[102]

In line with this, medicine has tended towards increasing centralisation and specialisation, with hospitals becoming fewer in number and requiring patients, staff, goods and visitors to travel ever further. In 2001 NHS staff, patients and visitors in England and Wales travelled an estimated 25 billion passenger kilometres, with car and van travel accounting for 83%. This does not include the 46 million annual outpatient attendances or the users of the 47,000 GP practices, nor

"The whole problem of health in soil, plant, animal, and man [is] one great subject."
– Sir Albert Howard

"The art of medicine consists in amusing the patient while nature cures the disease."
– Voltaire

pharmacies, dentists or opticians. In fact, the Department for Transport (DfT) has estimated that as much as 5% of all UK transport is generated by the NHS.[103]

There is a certain irony to this, as road traffic accidents are the leading cause of death for Europeans under 45, and it is estimated that for every death on our roads another 20 people are hospitalised and yet another 70 require outpatient medical treatment. Without even considering the effects of air pollution, the DfT estimates that deaths and injuries on our roads cost the NHS £470 million and the UK economy £18 billion every year (the total cost to the NHS in treating all personal injury cases *other than* road traffic accidents is estimated to be only £170 million to £190 million).[104]

The prevalence of driving is far from the only effect our current lifestyles have on our health. It is well-known that our poor diets, lack of exercise, high alcohol intake and levels of smoking also all contribute significantly to levels of ill health. For example, the link between heart disease and low levels of physical activity is well-established, and Cancer Research UK advises that around half of all cases of cancer diagnosed in the UK could be avoided through changes in lifestyle.[105]

So between our unhealthy lifestyles, rising energy usage, escalating energy costs and its financial worries, the NHS is already facing severe challenges, before we even consider what Dr Robin Stott has described as the "most significant public health problem of this century" – climate change.[106]

UK climate change will mean that heatwaves will become more common (there is a 1 in 40 chance that the south-east will experience a serious heatwave killing thousands in the next four years, and a 1 in 4 chance of it happening in the next nine years), that sunburn, skin cancer and poor air quality rates increase and that the risk of hurricanes and flooding increases, with all their many health consequences. Climate change may also cause an increase in infectious diseases in the UK, affecting food-borne, water-borne and vector-borne disease.[107]

The NHS is aware of these growing challenges, and is considering appropriate responses. The Department of Health commissioned a group of independent scientists to produce the report *Health Effects of Climate Change in the UK*, which was published in 2002. This was among the first of its kind in that it focused on the quantitative aspects of possible impacts of climate change on health, describing itself as a "first look at a difficult problem". An update to the report, published in 2008, found that:

"Key areas for the NHS in adapting to climate change include: adapting the health and social care infrastructure (hospitals, nursing homes) to be more resilient to the effects of heat, gales and floods; development of local 'Heatwave', 'Gale' and 'Flood' plans for coping with disasters; and increasing awareness of how people can adapt to changes in climate." [108]

The NHS also published its *Draft Carbon Reduction Strategy* in 2008, which opened its consultation on ideas for reducing its energy use and carbon footprint, noting that the NHS is both the biggest employer in Europe and the single largest public sector contributor to climate change (producing around 3% of England's emissions).[109]

Overall, then, we are facing a familiar diagnosis – as in the rest of our society, we need radical changes to the way our present system operates to respond to the future we can see coming. And again, within the broad stream of our cultural stories about the future, the stories we hold around health and well-being need to shift in order to create the public and political will for a different approach.

Cultural story change

This Transition Falmouth composting day provides exercise and strengthens community ties, as well as re-skilling

The cultural stories that we hold about health, life and death affect our lives in many ways,

but in a time of energy descent they become even more crucial. We have considered the challenges our healthcare paradigm is likely to face over the next twenty years, but we can at least look to an example of a successful healthcare system in comparable circumstances – that of Cuba.

Cuba today has similar life expectancy (78 years) and infant mortality rates (0.5%) to the UK, but uses a far lower-energy, lower-cost system. It is very much community-based, with a much higher doctor-patient ratio than ours providing a small surgery in each village, and giving doctors the ability to diagnose on the basis of the social and psychological factors affecting the patient, as well as physical symptoms. It is also based far more on preventing illness rather than treating it. Here in the UK we often hear laments for the old 'family doctors' who used to provide a similar service.[110]

One of our dominant cultural stories in this area is that more healthcare is always better and that the benefits of treatment always outweigh the costs. Yet it is this attitude that led to our current bloated, inefficient and vulnerable healthcare system, and to growing problems like 'antibiotic-resistant bacteria' (which are evolving rapidly due to our over-reliance on the limited number of antibiotics at our disposal[111]), polypharmacy (regularly taking three or more medications, thus greatly increasing the dangers of their little-understood interactions) and huge investments in keeping people alive at a very low quality of life, sometimes even when they wish to die.

It has been said that in our society we

"If you are going to deal with the issue of health in the modern world, you are going to have to deal with much absurdity. It is not clear, for example, why death should increasingly be looked upon as a curable disease, an abnormality by a society that increasingly looks upon life as insupportably painful and/or meaningless."
– Wendell Berry

"You cannot have well humans on a sick planet."
– Thomas Berry

"Through our unrestricted use of energy and resources in the health care industry, as well as our production of greenhouse gases, we are actually contributing to the ill-health of our planet and ensuring future suffering of the Earth's inhabitants."
– Dan Bednarz and Kristin Bradford, 'Medicine at the Crossroads of Energy and Global Warming', *Synthesis/ Regeneration*, Winter 2008

tend to walk backwards into death, refusing to acknowledge where we are headed, but it seems reasonable at least that those whose health and quality of life have diminished to the point where they actually wish to die should be allowed to, freeing up energy and resources for those who do not. Similarly, those of us who do not wish to get ill should be supported in trying to achieve that aim.

The Transition Vision – looking back from 2027

As the changes we have examined in previous sections of the Transition Vision played out, some of the major strains on our healthcare system reduced, with heart disease, obesity, cancer and traffic accidents all seeing steep declines between 2010 and 2025.

Nonetheless, the challenge of reducing energy consumption in line with our international commitments still meant a period of great upheaval in the NHS. Within a few months of adopting our national TEQs system to implement our emissions commitments under the '2010 Accord' (see p.33), our Government produced a sweeping set of health service reforms designed to reduce energy consumption and oil dependency in the NHS, building on and extending the work already begun by innovative Public Health practitioners and the *NHS Carbon Reduction Strategy*.[112]

What used to be known as 'alternative medicines' were embraced, as practices like herbalism, acupuncture, massage and osteopathy became core pillars of public healthcare, with a big investment in teaching these skills leading to a blossoming of independent regulated practitioners in most communities. Virtually all towns and many villages also now have their own practising pharmacy, based on the now-famous 'One Mile Pharmacopoeia' concept. Certain specialist medicines and experts still travel the length of the country, but the bulk of the medicines needed locally are created in labs behind the pharmacies using locally grown ingredients.[113]

Gathering ideas about health at the 'Unleashing' of Transition Town Brixton

This decentralisation was also seen with a return to community-based healthcare, which aims to prevent people ever needing to attend a hospital for extensive treatment. Increasing the number of doctors substantially was deemed a stretch too far for the NHS in difficult financial times, but numbers of nurse practitioners (registered nurses with special training for providing primary healthcare) have increased dramatically around the country, and they are often the first point of contact for patients, having taken on many of the functions traditionally performed by doctors. They also often play an important role in supporting local agriculture, spreading knowledge on safe methods of food preservation and testing soil for toxicity.

The shift from a 'sickness service' to a 'health service' also includes the popular free NHS provision of so-called PELFs – Personal advisers on Exercise Lifestyle and Fitness – who offer advice and encouragement to their local clients on improving health and beauty through diet and lifestyle, as well as noting the financial benefits of joining the national surge in walking and cycling (pp.73-77).

Mental health problems have also reduced, from directly affecting around 1 in 4 people's lives in 2007 to just over 1 in 6 people in 2026. While it is hard to determine direct causes for such developments, this is in part attributed to the increase in physical wellbeing in the population.[114]

Between 2015 and 2027 average life expectancy has increased from 80 to 82, but crucially the level of health people experience at a given age has also increased substantially. People are not only living longer but also feeling healthier and happier.

Chapter 12

Wildcards

Wildcards are events which by their nature are hard to put a timeframe on, but which could have dramatic impacts if they occur. Some possibilities are listed here in rough order of likelihood:

- **Wars** Wars fit our definition of a wildcard, but it is inevitable that war will be taking place in the world over the next 20 years, with our cultural stories heavily affecting the likelihood and scale of such conflicts. Many have argued that conflict over energy resources has been the key driver of war historically, and with climate change and peak oil only set to increase such tensions, our attempts to resolve our energy-supply challenges will also have great impact here. This was demonstrated in July 2008 when Gordon Brown's statement "We stand ready to give help to the Nigerians to deal with lawlessness that exists in this area and to achieve the levels of (oil) production that Nigeria is capable of" prompted the collapse of a ceasefire in the region. Nuclear war also remains a real possibility.[115]

- **Pandemic** If we look at a chart of human mortality over the 20th century the World Wars are minor events compared with the 1918 influenza pandemic, despite that strain of influenza causing only around 5% mortality. Current travel habits mean that an outbreak today would travel extremely rapidly.[116]

- **Nuclear disaster** on the scale of Chernobyl or worse.

- **The collapse of the United Nations** or a fundamental change in the nature of government in the UK. As the EU commented in 2008, "The multilateral system is at risk if the international community fails to address the threats (of climate change). Climate change impacts will fuel the politics of resentment between those most responsible for climate change and those most affected by it ... and drive political tension nationally and internationally." [117]

- **Transformative technological breakthrough** e.g. self-replicating nanofactories.

- **Collapse of the internet.**

- **Human sperm count or fertility collapse.**

- **Extraordinary natural disaster**, such as an asteroid or comet hitting Earth.

- **Extra-terrestrial or divine intervention.**

Chapter 13

An overview – systems thinking

'Systems thinking' considers the relationships between aspects of a system, rather than trying to isolate component parts and examine them separately. This perspective can help us in seeing the deeper questions of resilience underlying all of the disparate areas examined in Part Two.

In Rob Hopkins' *The Transition Handbook* he describes three features that make systems resilient to shocks – modularity, diversity and tightness of feedbacks.[118]

Modularity is the extent to which given parts of a system are equipped for a substantial degree of independence. How well placed are they to continue functioning if another part of the system is disrupted or disabled? How well could Britain feed itself in the case of shocks to the international finance markets . . .? How well could your household feed itself if fuel disruptions affected food deliveries to supermarkets. . . ?

Diversity follows on from this. In a modular system, each 'module' also needs to be different from the others to maximise resilience, otherwise a single shock might damage or disable all of the modules. As an example, individual elm trees can thrive outside of a forest (they are modular), but they proved too similar to resist the invasion of Dutch Elm disease. Ideally each module in a system will adapt to its local conditions and required functions and so provide both excellent functioning and a unique module for the system as a whole.

The modules in a given system might be people, species, machines, institutions, sources of food or whatever, but the greater the variety and the more possible responses they have to shocks the more resilient that system will be. A healthy and mature ecosystem is an example of a highly diverse system that is highly resilient to shocks. A tight-knit, innovative human community, well-adapted to its local environment and holding a wide range of skills, resources, stories and available responses is similarly better equipped to deal with change.

"Our ignorance is not so vast as our failure to use what we know."
– M. King Hubbert

"Not everything that counts can be counted. And not everything that can be counted, counts."
– Albert Einstein

In this context our third feature, *tightness of feedbacks,* means the 'reaction time' of the system – how quickly it responds to changes in the environment or in the system itself. Human-scale systems have the same short reaction time that individual humans do, but vast globalised systems can be much slower. It is akin to the difference in reaction time between a shoal of fish and a supertanker. There are circumstances in which a slow reaction time can be a benefit, but generally speaking, faster reaction times increase resilience.

So, high modularity and diversity and tight feedbacks are all important to the resilience of a system. Unfortunately, the dominant story shaping the systems in our culture recently has been that of *cost-efficiency,* which identifies a single desirable outcome and tries to achieve it as cheaply as possible by removing all other 'redundant' elements. In other words, prioritising cost-efficiency tends to involve removing all possible modularity and diversity from systems, meaning that when an element of the system fails for any reason, there is no replacement or alternative ready to take its place. This can bring huge interdependent systems grinding to a halt until that one element can be repaired or replaced.

Image from www.imageafter.com

Per Bak, in his book *How Nature Works,* outlines a nice way to demonstrate these principles in action. He asks us to imagine building a sand pile by continually adding grains of sand. As we continue adding more grains, we soon observe that avalanches of sand start occurring, seemingly at random.[119]

Yet it is not random. In reality there is a critical level of steepness beyond which grains of sand break away and start tumbling down the pile. Now imagine that, as we add sand, all the areas which reach this critical level of steepness are marked red, so that any sand dropped on a red area will cause an immediate avalanche.

We notice that the red patches first appear as tendrils running down the side of the pile, but as we continue to add sand (carefully avoiding the red areas) the pile gets steeper and more little tendrils of red appear. Eventually we observe that the tendrils of red start to join up, and discover that if we trigger an avalanche here, all the connected red areas are also drawn in, creating a bigger avalanche. Alternatively, if we drop a grain onto an isolated red area, then the avalanche will be confined to that one red tendril running down the side of the pile.

This is the principle of *modularity* in action. The more *inter*dependent the areas of the pile are, the more vulnerable they are to a simultaneous collapse. Clearly the *diversity* of the elements in a pile of sand is also fairly low! The '*reaction time*' in this case would be how quickly we adjust our aim to avoid dropping grains of sand on the spreading red areas.

By carefully only dropping sand on areas that are not yet red, we are able to delay experiencing 'failure' (avalanche) in any part of the system, but the trade-off is that the red areas go on spreading. This means that when we finally run out of alternatives and are forced to drop sand on a red area (because the whole sand pile is now in this critical condition) there is a huge avalanche – *every part of the system fails simultaneously.*

What has happened is that our quick reaction time has allowed us to effectively buy time by selling off modularity step-by-step until we had no more to sell. By doing so we have successfully delayed the avalanche, but also made the inevitable eventual avalanche much more severe. We might have been better off allowing a series of smaller avalanches which could be relatively easily contained, unless we are able to somehow stop the input of new grains of sand, or change the critical steepness threshold.

At this point we can draw the analogy to our complex human systems. The continual input of sand is perhaps analogous to the increasing load on human systems caused by our growing population and growing levels of debt.

The obvious broader analogy here is in the financial world, where numerous complicated financial instruments are designed to share out risk across economic systems, thus effectively decreasing modularity. The consequence of this is that while collapses (bankruptcies) may be made less common, the collapse of one large company or financial institution can threaten to bring down whole interdependent swathes of large economies, rather than merely affecting that company's staff.

This is why the Green New Deal Group effectively argue for increased modularity, saying that:

> "Instead of (financial) institutions that are 'too big to fail', we need institutions that are small enough to fail without creating problems for depositors and the wider public." [120]

In his 2008 article 'The Failure of Networked Systems', David Clarke insightfully extends the analogy to our energy challenges. He highlights that as high quality oil ('sweet light crude') has become more difficult and expensive to extract we have put off the consequences of that by bringing in alternatives like Liquefied Natural Gas (LNG) and biofuels, and argues, using the sand analogy we have just examined, that,

> "Essentially the part of the network called 'Sweet Light Crude' turned red, so we started connecting the 'Oil Network' to other networks." [121]

We started reducing modularity by linking oil to natural gas, and then used biofuels to link these to the food and energy markets as well, meaning that problems in any of these areas increasingly cause problems for the others. Some of the alternative energy sources we are turning to also produce significant greenhouse gas emissions, so the sustainability of our entire global environment is also connected to the energy/fuel system as it struggles to deal with its growing load.

As with our sand pile example, we have been selling off modularity in order to put off the consequences of an ever-increasing load, but does that mean we are headed for a 'big avalanche' – a widespread breakdown in many of our increasingly interdependent human systems?

If we cannot stop the 'flow of sand' – the growth in consumption, and attendant growth in ecological and economical debt – then the answer appears to be yes. This is why our cultural stories hold such importance at this point in history – if we do not change the underlying paradigm of growth then we will face consequences that we can see with ever-greater clarity. Yet if we can change them, there is still a window of opportunity to face and address our 'debt problems' and move to living sustainably within the carrying capacity of our planet's ecosystem – the great overarching system that has taught us everything we know about resilience.

If we want to bring our Transition Vision of the future into reality, it is clear that we must collectively recognise and respect the limits on available fossil-fuels and on our atmosphere's ability to absorb greenhouse gas emissions. It is economics that currently acts as the monitoring and communications hub of our societal system – deciding where and when we should act – so we must link our economic systems to this concept of *limits*.

The finite system of which we are all a part.

This is why TEQs provide such a crucial aspect of the vision, as they provide the economic framework within which the diversity of small-scale solutions in Transition Towns and other communities can flourish. TEQs would enshrine this recognition of environmental limits in law, so that all the systems (communities, companies etc.) within the UK can rest assured that the large-scale problems of climate change and peak oil are being faced, leaving them empowered to find their own creative solutions for thriving within that context and united in a co-operative common purpose to do so. [122]

Of course we should not pretend that limiting economic growth will be painless, but it is inevitable. It is true by definition that all life on this planet *will* ultimately live within the ecological limits of our environment. Temporary overshoot is possible, but limits are limits.

Any system that depends on continual growth is doomed, and the earlier we acknowledge this reality and act on it the less painful it will be, like that trip to the dentist that we keep putting off for just one more year.

Our economic systems must recognise that maximising growth in Gross Domestic Product (GDP – the measure of economic growth) is not the be-all and end-all of human existence, and that in fact many activities that increase GDP actually decrease our quality of life through destructive impacts on our true assets – nature, people, social capital and our built infrastructure.[123]

For many of us, discussions of economics have appeared uninteresting and confusing, seeming to have little to do with what really matters to us. This is actually an important insight, but by not engaging with the discussion we have allowed it to be lost, leaving misconceived systems in place to drain the true wealth that supports all our lives. By relocalising our economies within a system like TEQs we can begin to free ourselves from this burden and get back to focusing on the human-scale activities and relationships that shape our day-to-day lives.

"The economy is a wholly owned subsidiary of the environment, not the reverse."
– Herman Daly

"We act as though comfort and luxury were the chief requirements of life, when all that we need to make us really happy is something to be enthusiastic about."
– Charles Kingsley

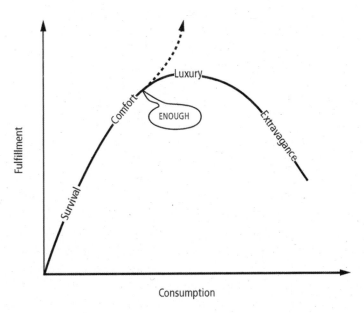

Figure 10: Consumption and fulfilment.[124] **Based on Dominguez and Robin (1992)**

Part Three

MAKING BEST USE
OF THIS TIMELINE
by Rob Hopkins

This book is designed to be of use to those planning for the future of their communities, in particular to those engaged in using the Transition model.

Although the Timeline outlined previously in this book is useful in its own right, for example in providing Transition initiatives with a new and clear narrative to communicate, it is when a Transition initiative reaches the last of the 12 Steps of Transition that it really comes into its own.

Step 12 is 'Create an Energy Descent Action Plan', and what follows is a speculative look at the role Timelines can play in underpinning and informing the work of Transition groups at this most vital stage of the local resilience-building process.

The 12 steps are not prescriptions, but principles that anyone can use to begin the process of transition from oil dependency to local resilience. It is worth noting that although the term 'Transition Town' has stuck, what we are talking about are Transition Suburbs, Transition Islands, Transition Valleys, Transition Anywhere-You-Find-People. There are now many hundreds of these groups and organisations across the world, grappling with the very real issues of trying to design pathways their community might take in order to safely descend the energy mountain.

"Once you have glimpsed the world as it might be, as it ought to be, as it's going to be (however that vision appears to you), it is impossible to live compliant and complacent anymore in the world as it is."
– Victoria Safford

"What would happen if one day when we wake up, we realise that we are the majority?"
– Mario Benedetti

The 12 steps of transition

1. **Set up a steering group and design its demise from the outset.** This stage puts a core team in place to drive the project forward during the initial phases.

2. **Awareness-raising.** Build crucial networks and prepare the community in general for the launch of your Transition initiative.

3. **Lay the foundations.** This stage is about networking with existing groups and activists.

4. **Organise a Great Unleashing.** This stage creates a memorable milestone to mark the project's 'coming of age'.

5. **Form sub-groups.** Tapping into the collective genius of the community, for solutions that will form the backbone of the Energy Descent Action Plan.

6. **Use Open Space.** We've found Open Space Technology to be a highly effective approach to running meetings for Transition Town initiatives.

7. **Develop visible practical manifestations of the project.** It is essential that you avoid any sense that your project is just a talking shop where people sit around and draw up wish lists.

8. **Facilitate the Great Reskilling.** Give people a powerful realisation of their own ability to solve problems, to achieve practical results and to work co-operatively alongside other people.

9. **Build a bridge to Local Government.** Your Energy Descent Action Plan will not progress too far unless you have cultivated a positive and productive relationship with your local authority.

10. **Honour the elders.** Engage with those who directly remember the transition to the age of cheap oil.

11. **Let it go where it wants to go . . .** If you try and hold onto a rigid vision, it will begin to sap your energy and appear to stall.

12. **Create an Energy Descent Action Plan.** Each sub-group will have been focusing on practical actions to increase community resilience and reduce the carbon footprint.

Chapter 14

Timelines and energy descent plans

What is an Energy Descent Action Plan (EDAP)?

As this book has stressed, we are entering times of extraordinarily rapid change. We are moving from a time when our economic success and our very sense of personal prowess and well-being were linked directly to our degree of oil consumption, to one where our oil dependency directly equates to our degree of vulnerability. It is necessary to urgently design and implement low carbon, resilient ways of living – climate change makes this imperative, peak oil makes it inevitable, and Transition initiatives make it feasible and, hopefully, desirable.

When our local authorities sit down to create Development Plans for our communities, they invariably start with assuming a graph where all the lines rise as they move from left to right – more economic growth, more energy availability, more housing demand, more traffic. All of these are now seriously in question, yet little creative thinking is going on with regards to how to plan for this new context.

The concept of creating a 'Transition

Vision' for our communities is something that becomes increasingly urgent with each day that passes. This, in effect, is what an Energy Descent Plan is. However, it is more than just a dry planning document full of maps and noble yet indistinct aspirations to sustainability. It is, as much as anything, a new story for the community. By starting with a vision of a post-oil, low-energy world as a happier, more fulfilled, slower, more satisfying place, an EDAP then works

edaptation // noun.
1. the act of a community adapting to climate change and peak oil by creating and actioning an Energy Descent Action Plan.
2. the act of transitioning to a post-carbon future in a positive, proactive manner
3. (sociology): a far-reaching and widespread yet significant modification of individual and community attitude and behaviour.
4. the act of moving from oil dependency to local resilience community by community.
5. the act of personal attitudinal and behavioural change.
From the Sunshine Coast EDAP

"It is so encouraging that a council that has set a challenging goal to become Australia's most sustainable region has a community that is leading the charge for a sustainable future. The Energy Descent Action Plan is an inspiring and innovative document – only the second in the world to be completed – and will provide a valuable contribution to Council to inform our planning to address some of the most difficult issues of our time."
– Cllr. Keryn Jones, Sunshine Coast Council

"I am looking forward to receiving the Energy Descent Plan from Transition Sunshine Coast. Climate change is the most important issue that all levels of government need to tackle with vigour, intellect and commitment. We will only succeed in our goals of reducing greenhouse gas emissions by working in partnership with our community. I congratulate Transition Sunshine Coast on their pro-active work on this key issue. We are blessed on the Sunshine Coast of Australia to have such a visionary and energetic group committed to tackling the biggest issue of our times."
– Vivien Griffin, Division 9 Councillor, Sunshine Coast Regional Council

backwards, working out how practically to get there over a 20-year period.

An EDAP should be a celebration of the creativity of the community, weaving together artwork, 'Transition Tales', local history, the practicalities of moving away from oil addiction, and much more. We often stress in Transition that we need to create visions of a post-carbon world so enticing, so compelling and attractive that people leap out of bed in the morning determined to dedicate their lives to its implementation. An EDAP is an embodiment of this.

EDAPs so far

If you are reading this thinking that an EDAP is a fully tried-and-tested approach, with many years experience of being implemented around the world, what follows is an Important but Cheerful Disclaimer. This is a new approach with an evolving set of tools, and is really, as with the Transition approach itself, being made up as we go along. Although there is not a huge body of examples to draw from, there are one or two that you might find useful. The first EDAP was created in Kinsale in Ireland, by second year permaculture students at the Further Education College. It was a spontaneous response to finding out about peak oil, and it was created largely as a response to being unable to find other examples of communities thinking ahead and planning for peak oil.

The Kinsale EDAP[125] is a fascinating exploration of the potential of taking a proactive response to resource depletion and climate change, and became an extraordinarily viral phenomenon, being downloaded many thousands of times, with many people finding that it was a missing piece of the jigsaw puzzle that they had been looking for for years. It is important to observe though that it was largely a student project, not grounded in extensive community consultation or founded on a deep process of involvement and engagement.

The first really thorough EDAP was developed on the Sunshine Coast in Queensland, Australia, by Sonya Wallace, Janet Millington and 22 course applicants at Transition Sunshine Coast, the first formal Transition initiative in the country. It grew out of the 'Time for an Oil Change' course run at the Sunshine Coast Energy Action Centre, which explored what the model developed in Kinsale might look like if applied to the Sunshine Coast. The Sunshine Coast EDAP is a bold and pioneering example of people using the Transition principles in processes that are rooted in their community. Its recommendations include, among other things, the need for a food self-sufficiency strategy for the area, exploring the role of forestry and bamboo and the need for training and education across the area. Although still in draft form at the time this book went to print, it is an exciting and powerful demonstration of the EDAP concept in practice.

Sonya Wallace (left) and Janet Millington unveil the Transition Sunshine Coast EDAP in Queensland, Australia

The Totnes EDAP

Totnes was the first Transition initiative after Kinsale, and after two years of work, it began to develop its EDAP in mid-2008. Funding from the Esmée Fairbairn Foundation and from Artists Project Earth mean that the Totnes EDAP process is the first with some funding resource, and the hope is not just to create an EDAP, but also to identify the process for creating one so as to enable their creation to spread widely.

Transition Town Totnes (TTT) began in 2006 with its 'Unleashing', and since then its spread through the community can be described as being like mycorrhiza, with its various working groups (Energy, Food, Building etc.), doing a great deal of thinking about how to build resilience in their particular field of the Transition process, as well as initiating a wealth of projects, initiatives and connections.

The Totnes Energy Descent Pathways project is about pulling those various strands and threads together to create a plan of action for Totnes & District, as well as for the future of TTT itself. The project was launched in September 2008 and it is planned that the Plan itself will be published in June 2009. It has been a process that has made use of a great deal of public consultation as well as using many innovative public engagement tools (some of which are described in *The Transition Handbook*).

The role of the Timeline for EDAP teams

Our first experiments with the idea of creating a timeline came about in one of those creative brainstorming hours where ideas seem to flow very easily. As part of preparing for an event at which people would be invited to write stories from the future, a small group of us sat down to try and create a 20-year Timeline for Totnes, beginning by marking in some of the events that we thought were likely, such as peak oil happening in 2010, and then we began to weave events and developments around that, from the practical to the absurd. The resultant timeline became a great talking point at events, and was usually surrounded by people in rapt attention, studying it closely.

For the more thorough and in-depth Energy Descent Pathways project, a more robust and adaptable version was created. Large card sheets were used, and fixed together so it could all fold out into an 8-metre-long timeline. The years were marked onto it, and the whole thing covered with a protective covering to keep it clean. At events people are

"If you want to go fast, go alone. If you want to go far, go together."
– African proverb

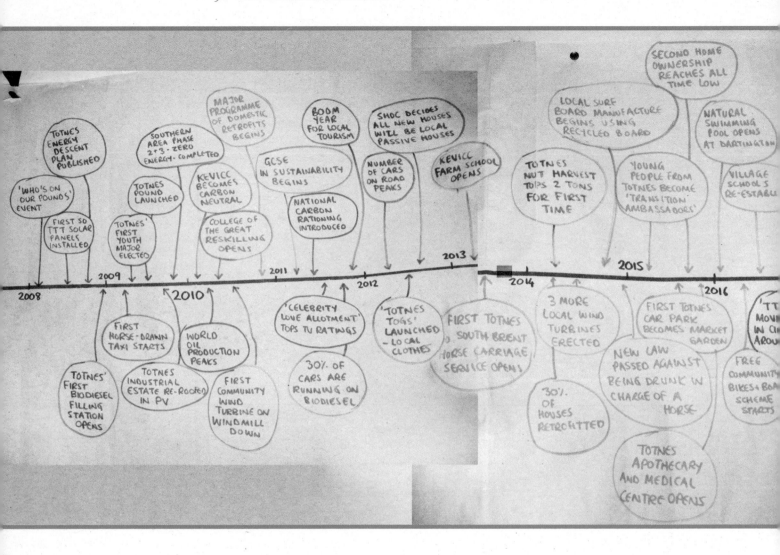

invited to write their contributions onto Post-It notes and put them at the relevant year. As the project evolves, more and more Post-Its will be added to the timeline, building up a fascinating resource, eventually several layers of Post-Its thick, resembling a shaggy animal rather than a pinboard. At the culmination of the project, and in order to ensure that the whole process is well-synchronised, a master timeline will be created, with all the different strands of the Plan being overlaid so as to create a coherent narrative.

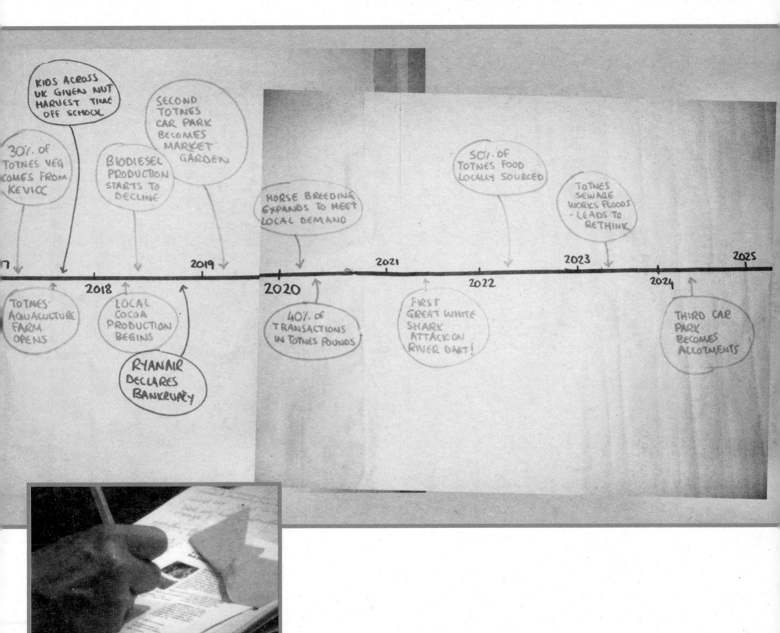

The Palette of Tools

The Totnes EDAP process makes use of a broad range of tools. Some of these are set out below in a toolbag of practical exercises that you can take and use in your community. Many of them are designed to facilitate the generation of Timelines, or to run alongside their use, and they all strive to be inclusive, fun and engaging. We are always keen to hear of new tools and activities that you create or find to be useful. The tools set out below are:

1. The 2030 School Reunion

2. Backcasting

3. Transition Tales (with adults and children)

4. Visualising the Future

5. Resilience Indicators

6. The 'EDAP in 2 Hours' exercise

Timeline tools #1. The 2030 School Reunion: a powerful tool for visualising the future

This exercise was first performed at the launch of the Totnes Energy Descent Plan in September 2008. The premise is that it is 2030, and four residents of the town are meeting for a school reunion. Each of them has seen great changes in their lives, and they have a lot of catching up to do.

It opens with a narrator reading an account of the intervening 20 years, in which the town experienced great difficulties, as economic contraction and the end of cheap energy led to economic upheaval. Many businesses closed and things became very difficult for people. Over that time, however, a counter trend began to emerge very strongly, the trend towards rebuilding a localised and diverse economy. On people's seats are cards, set out in advance, which contain different facts about each character.

The audience gathers into four groups, with one character (played by an actor dressed as that person) and a facilitator in each group. Each character tells their life story about their experiences during these times of extraordinary transition, with the audience questioning them and adding details about their lives. Together they create the character's life story which will be shared at the School Reunion. This is a very engaging part of the process.

Then the four actors stage their meeting at the school reunion and chat about the last 22 years. If properly done, this can be alternately moving, hilarious and absorbing. It works best if the actors have planned the session in advance, but have still left enough room in to improvise and to interact with the audience.

This is an exercise which requires a fair degree of forward planning, and some thinking about in advance. You can find all the notes and information you will need to do this exercise at: www.tinyurl.com/6mxz4t.

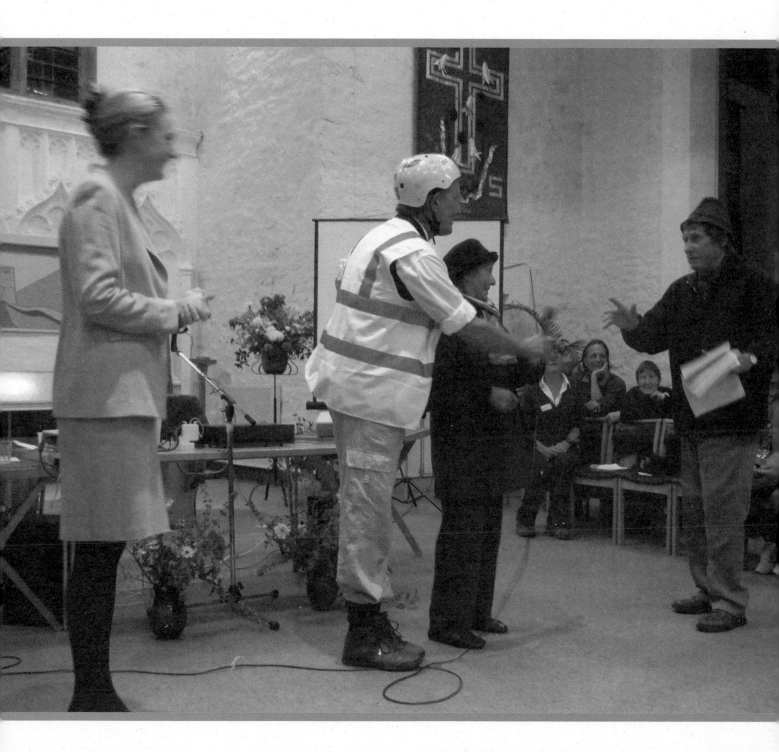

Timeline tools #2. Backcasting

As discussed in Part One, if we just sit back and allow the next 20 years to unfold, responding reactively to events, the outcome can be guaranteed to be disastrous. Peak oil and climate change demand a decisive, co-ordinated and thought-through response, and not just at one level – we need an interlocking set of responses at the local, national and international levels. As this book demonstrates, the Transition approach is to start with a vision of how the future could be, and to work backwards.

A practical example

Let's say we decide that by 2015, any new house built in your community has to be a 'local passive house', that is a house built to passive house standard (i.e. so well designed, oriented and insulated that it can be heated from your body heat and your daily activities) and built using 80% local materials (clay, straw, timber, hemp etc.). That means that by 2013 the local planning department needs to be familiar with the concept, and local builders need to be retraining in how to use these materials. By 2012, local farmers need to have started growing hemp in order to feel confident in growing it, and they need to be forming co-operatives to enable them to purchase the equipment to process the hemp. By 2011, the first prototype building using this technology needs to have been built so that it can be studied and lessons learnt for its wider roll-out. And so on. It is only by backcasting that one can see the context of a particular activity and the route to ensuring that all the elements that will maximise its chance of success are in place.

Resources:
James, S. & Lahti, T. (2004) *The Natural Step for Communities: how cities and towns can change to sustainable practices*. New Society Publishers. An excellent overview of a process similar to Transition which has been run very successfully with communities, with some great case studies outlined here, mainly in Scandinavian countries.

Cook, D. (2004) *The Natural Step: Towards A Sustainable Society* (Schumacher Briefing No. 11). Green Books. A more theoretical overview of the Natural Step concept.

(Right) Backcasting work using a timeline, at Transition Sandpoint, Idaho.

Timeline tools #3. Transition Tales

Transition Tales can be done with adults and children, indeed some adult ones appear in Part Two of this book. What follows is a report which Hannah Mulder of Transition Town Totnes wrote for TransitionCulture.org, describing the work her group did in the local secondary school.

"In the Autumn of 2006 a group started to look at how we might introduce the ideas of transition to young people and using storytelling from the future as a format was settled on. We set up a meeting with local secondary and primary schools at Schumacher College to talk to them about what they felt would work and was needed, and from that we were invited to do a pilot with a Year 9 class (13-14 year olds), at King Edward Sixth Community College (KEVICC), the big local secondary school.

"Having thought that our end result would be written stories, we found that the format of filming local news stories from 2030 worked really well and KEVICC then invited us to work with the whole of Year 7, (11-12 year olds), in April and May 2008. This was to be part of their 'collapsed curriculum' year in which students would not study under the usual headings but would look at overarching topics, such as 'change', which is where we came in.

"Having secured funding from The Funding Network, we were ready to roll and started to fine tune the workshops and gather the resources that we'd need. A pilot with a Year 8 group helped us to develop the workshops and during the thirty-three sessions with the Year 7 group we tried out various different ideas, exercises and formats, especially in the middle bridging session.

The workshops

"Our structure at KEVICC was three two-hour workshops, but we've also run one-hour sessions at primary level (very short!). We introduced the issues through various games and exercises during the first session, and in the last session created and filmed imaginary local news stories from 2030, which we then posted on YouTube so that the students could share them with their friends. In the middle session we tried out a variety of things.

"One session was about getting out into nature, sitting around a fire and hearing a resonant story, interacting with and being present in a wild place; another session developed a 'Quest' game in which the passions, skills and resources of the students were taken forward into the future to meet various challenges. In our final trial middle session we looked at the school community, envisioning through three-dimensional map-making how the school could be transformed in the future. The obstacles to making those changes happen were then explored and the students completed the session by making superheroes out of potato, symbolising the qualities needed to overcome those obstacles!

"At the end of the sessions we had some glowing feedback from the school itself. Emma Osbourne, Year 7 Curriculum Co-ordinator, wrote:

'I have really enjoyed working with the Transition Tales project. I feel that this is incredibly worthwhile. We spend a lot of time in education worrying about targets and levels. It has been wonderful to be involved in helping young people to think realistically about their future. People often talk about educating students about these issues and usually this just means preaching at them in an assembly or having a one-off awareness day. The potential for this to be ongoing and the project's emphasis on this is really encouraging. The approach you and your team have taken is fantastic. I like the way you have focused on the opportunities created by this inevitable change and how we can take control of them. It wasn't at all preachy or scaremongering.'"

Resources

The films from the Transition Tales project at King Edward VI Community College can be seen at: www.tinyurl.com/5gqyap

Transition Tales with Adults.

Telling stories from the future is such fun that it ought not be reserved just for children! In Totnes, we have run several workshops with adults, the most recent of which was with the local Wondermentalist Cabaret, a collective of writers, poets and performers. During the daytime workshop the group wrote pieces around 'A Day in the Life in 2030', looking back from 2030 to now, 'offered and wanted' adverts from the local paper, and also taking that week's local paper and writing stories from 2030 using the headlines. The material generated was then performed at a Cabaret that evening, and mixed the hilarious with the poignant and the thought-provoking.

Resources

The films from the Transition Tales project at King Edward VI Community College can be seen at: www.tinyurl.com/5gqyap, and the podcast of the Wondermentalist/Transition Cabaret can be heard at www.tinyurl.com/dzdl8r.

Timeline tools #4. Visualising the Future

For many people, being able to imagine the post-oil world is a powerful step towards being able to believe that it is not only desirable but possible. Putting yourself, mentally, in the future, can lead to some surprising insights. Different Transition groups and practitioners have developed tools for Transition visualisations. Here are a couple of them.

The first was developed by author and activist Starhawk, and was transcribed from a talk she gave in Totnes. After an initial invitation to get comfortable, to take a few deep breaths, she continued:

"I'd like to invite you to visualise yourself walking down Totnes High Street in a future Totnes, a future which has successfully navigated the energy descent transition. As you walk along, look around you. Who are you? What are you doing? Look at the shop fronts and the buildings, what has changed? What are the businesses you see? Sniff the air, how does it smell?

What did you have for breakfast? Where do you get your food from? Who else is walking with you? Are there children there? Who cares for them? Who teaches them so they know what they need to know? Are there old people? Who cares for them? How do they share their wisdom? What kind of work do you do in this future? What are your rituals and celebrations?

Imagine someone coming towards you, someone from this future time. You sense they have a lot of knowledge, wisdom and memory. What do they look like for you? They have the knowledge and the memory of how this Transition came to be. Talk and listen to them. Maybe you have some questions for them, maybe they have some advice for you.

[here she left a gap for people to reflect on this. . .]

Know that you may not get all the answers this second but you can invite this future to inform your dreams, your choices and allow it to inform your bringing this future into being. Now, as you start to leave, you walk back up the High Street, and as you walk, notice what you pass . . . what has changed, and what remains the same. Bring that memory and that vision with you, so that we can create the deep knowing that it is possible and that you have an important role to play."

When she had finished she talked about how at the moment, statues are dedicated to famous people. In this future, she said, the statues will instead be to the amazing work that many thousands of ordinary people put into our getting to that point. She invited people to start thinking about what they would like their statues to be like, what clothes they would like to be wearing in them!

Visioning a Positive Future: an exercise developed by Hermione Elliott for Transition Town Lewes [126]

The set-up

This is an interactive imagery process and the key is that we actively embody our vision of the future . . . we enter into it as if it were happening now. Remind people that this is a faculty we all have – as children in creative play, we moved between past, present and future very freely – we can approach this as an adventure where we might just discover something new and surprising.

Capturing information is important. If a good quality recording device is available, use that; if not, have two pairs of people act as scribes, recording the comments on flip-chart paper (change pairs halfway through the exercise). Divide the room into two halves using some masking tape on the floor to separate the time zones. Pin up a large notice with the present year e.g. 2009 at one end and a future year at the other e.g. 2030.

Explain that we are going to explore two possible futures – one where we aren't happy about the outcome and one where we feel good. We do this because we can often learn more from our mistakes than our achievements. Leave time between questions for people to reflect and ask them to speak up and share their insights. To begin, invite the group to stand in the present and start the exercise:

The present

"Picture yourself standing in the centre of your town or in your own home or garden. Experience it fully and notice how it feels. What's the atmosphere? What sounds are you aware of? What's the best thing about it? What's the worst? What do you fear? What do you hope for?"

Now we are going to go forward in time into a future where things haven't turned out as well as we hoped. Invite everyone to walk forward and move into the future time.

Negative future

"Picture yourself standing in the centre of your town or in your own home or garden. Imagine it's 2020 and we are feeling bad. Things haven't worked out at all well and life is hard. We didn't prepare well enough for reducing oil supplies or for climate change. We ignored all the warning signs and carried on as if nothing was happening. What does the result look like?

"Find out what is happening, where are you, what is the bad feeling. . . .what is the most important thing that makes you feel this bad. . . . How are people relating to each other. . . how are we managing our daily lives. . . where did we go wrong. . . if you were able to advise your younger self how to avoid feeling like this. . . what would you say. . . make sure the younger you gets the point about what they need to do in order to avoid feeling this bad."

Invite everyone to come back to the present (e.g. 2009) end of the room, to step back into themselves, to breathe in being back in the present, and to hear the advice from the negative future.

Now we are going to go forward in time to a future where things have turned out really well. Invite everyone to walk forward and move into the future time.

Positive future

"Picture yourself standing in the centre of your town or in your own home or garden. Imagine it's 2020 and we are feeling good. Things have worked out much better than we had hoped and there have been some happy surprises.

"Now [. . .] is a successful Transition Town; we have responded positively to the challenges and changes. Notice what life is like for you at this time in the future . . . what is going on around you. . . This is a time when you feel good about life here. . . what is this good feeling. . . what is the most important thing that makes you feel this good. . .

"As a successful Transition Town we are managing really well with much less oil. What has changed? How do we get around? These days, in the winter how do we heat our homes? What happens if we are sick? What do our schools teach? How do the streets look and feel? What is happening in our gardens and neighbourhoods?

"In getting to this point what have you learned the hard way.. what did you do to make this possible? If you look back to your younger self what message would you send that could help them react positively to the changes ahead?

"How is this different from the negative future we looked at before? Be as specific as possible in comparing the two."

Other possible questions could refer to food, waste, sources of energy, taking care of ourselves physically, emotionally, spiritually.

Invite everyone to come back to the present (e.g. 2009) end of the room, to step back into themselves, to breathe in being back in the present, and to hear the advice and insights from the positive future. Share in pairs and in the light of the exercise make three commitments that will help to create the positive future that you have experienced.

To finish, everyone join together to write or draw on large Post-It notes key elements of the vision of a positive future and to place them on a very large copy of a map of the town.

Exercise developed by Hermione Elliott for Transition Town Lewes, inspired by the book *Life Choices and Life Changes through Imagework* by Dina Glouberman.

Resources

Some useful future visualisation exercises can be found in Macy, J. & Brown, M. (1998) *Coming Back to Life: practices to reconnect our lives, our world.* New Society Publishers.

As well as this book, other useful documents which paint descriptive pictures of future scenarios include David Holmgren's *Future Scenarios*, Foresight's *Intelligent Future Infrastructure: the scenarios towards 2055* and FEASTA's *Energy Scenarios Ireland*.[127]

Timeline tools #5. Resilience Indicators

Resilience is the ability of a system or community to withstand impacts from outside. An indicator is a way of measuring that, a measure of collective robustness, if you like. Conventionally, the principle way of measuring a reducing carbon footprint is CO_2 emissions. However, we firmly believe that cutting carbon while failing to build resilience is an insufficient response to peak oil and climate change.

So what constitutes a resilience indicator? A resilience indicator is something that allows us to get a sense of whether or not the community is moving towards or away from resilience, and is something that can be returned to over time and checked to see if we are moving in the right direction. These might be things like:

- the percentage of new buildings which meet passive house standards

- the percentage of food consumed in the area that was grown in the area

- the amount of people who feel they have certain skills

- people's level of optimism that change is possible

- number of people who are active members of car-share clubs

A recent brainstorm in Transition Town Lewes identified the potential of also having indicators others might consider somewhat sillier, such as the number of chickens in the town or the number of people carrying non-plastic bags. What matters is that they are indicators that people can feel a degree of ownership over and which can be returned to over time to assess the direction in which the town is moving.

Getting started

The best place to start is with thinking about, in the context of your group, what the Characteristics of a Resilient Community might be for your area. Some examples might include:

- 'the community has a strategy for increasing local ownership'

- 'there is a strong belief in education at all levels'

- 'the community is able to create a vibrant network of local food production within a short time frame'.

They are written as a statement of fact from the perspective of the resilient community of the future. They pin down the question 'if you woke up in [insert name of your community] in 2030 and it had successfully Transitioned, what would it look like, feel like, smell like, sound like?' Can you pin down 5-10 characteristics of a resilient [ditto]?

There are three ways you could assess your indicator – through a survey or questionnaire of some kind, through research or through interviews with individuals. Developing a set of resilience indicators as part of your EDAP work means you can include them in your plan, which will make it much more robust.

Resources

Some of the most detailed work on the idea of Resilience Indicators comes from the Centre for Community Enterprise's *Community Resilience Manual: a resource for rural recovery and renewal* and the accompanying *Tools & Techniques for Community Recovery & Renewal* (available from www.cedworks.com).

(Right) Might the 'amount of local apples harvested' become a useful Resilience Indicator?

Timeline tools #6. 'The 2-Hour Energy Descent Plan for Transition Town Anywhere'

This is a wonderful process created by Lucy Neal and Duncan Law of Transitions Tooting and Brixton respectively in London. The idea is wildly ambitious. It is to offer a deep experiential immersion in the 12 Steps of Transition and also to actually create an EDAP in two hours! It does so through a process which is theatrical and fun, and which changes what it is doing so often that boredom is impossible. It was first 'performed' on the South Bank in London one afternoon in July 2008 as part of the LIFT Festival there.

The workshop involves the use of oversized props, such as a huge book (which becomes the EDAP) and a very large ball of string (which serves no obvious purpose). It features, in two hours, a peak oil talk, an Unleashing complete with cake and streamers, a talk from a local 'elder', and ends up in the writing of an EDAP! It is by turns thought-provoking, hilarious, engaging, illuminating and deeply touching.

The workshop not only enables people working on and interested in Transition in the place where the workshop is run to meet and find ways to work together and support each other, but also acts as a vibrant and inspiring introduction to the concept of Transition and a taster of the element of playfulness and enthusiasm driving it.

Resources

You can download a film of the LIFT Festival workshop at: http://tinyurl.com/5zuewc. The 'how to' instructions are long and detailed and would take up too much space in this publication, so we've put them online too at: www.tinyurl.com/6mxz4t.

Speculation on future tools

The tools outlined above are just a handful of those that could be used in the process of designing successful pathways away from energy-dependency for our communities. There are many more that would be very useful, and what follows is a speculative list of new tools that could yet be developed. These could include:

Computerised maps

Systems such as GIS mapping (Geographical Information Systems) allow overlaps of economic, social, ecological, geographical and other forms of data. Specifically designed GIS systems could allow communities to identify, for example, how much land would be needed to feed them, how much potential urban growing space there is, and so on.

Conflict resolution tools

It is not a given that once a draft plan is presented, it will meet with universal agreement. Some may object to certain energy proposals, or to other aspects of the plan. Those forming the ideas will also need to have the skills to manage any conflict that may arise.

Representativeness

Creating an EDAP is a powerful opportunity to form a powerful coalition of individuals and organisations, one that reflects the diversity of the community, as, after all, it is ultimately a plan for their future. New tools are needed to ensure that this degree of representativeness is embedded through the whole process.

Accessing useful data

An EDAP is far more robust if it is underpinned by firm data about the settlement in question, so firm guidance as to where to find this and how to make use of it would be very helpful. What information does the local authority hold? What can you get from local businesses?

. . . and hundreds of other things no-one has thought of yet!

How about plays designed around Timelines? Novels? A community-produced Timeline quilt? Art installations? Well-designed collaborative online tools to enable the generation of stories to weave through Timelines? Clever ways of ensuring that Timelines are continually adapted and responsive to events? Over to you!

"At first people refuse to believe that a strange new thing can be done. Then they begin to hope it can be done. They see it can be done. Then it is done and all the world wonders why it was not done centuries ago."
– Francis Hodgson Burnett

2010 2011 2012 201

WORLD OIL PRODUCTION PEAKS

Holistic Science Education in Primary Schools

HERN REA GE 2+3 D ENERGY PLETED.

WINDMILLS

SOUTH HAMS COUNCIL BEGINS TO...

GCSE IN SUSTAINABILITY BEGINS

CELEBRITY LOVE ALLOTMENT TOPS TV RATINGS.

1st TOTNES CAR-PARK BECOMES MARKET GARDEN

BOOM YEAR FOR LOCAL TOURISM

30% OF CARS ARE RUNNING ON BIO-DIESEL

THE PLANET SHIFTS POLES

MAJOR PROGRAMME OF DOMESTIC RETROFITS BEGINS

NUMBER OF CARS ON ROAD PEAKS

TOTNES TOGS LAUCHED - LOCAL CLOTHES

KEVICC BECOMES CARBON NEUTRAL

COLLEGE OF THE GREAT RE-SKILLING OPENS

MORRISONS TAKEN OVER AS INDOOR...

Part Four

GLOBAL CONTEXT – CLIMATE CHANGE/FUEL DEPLETION

This book considers the UK outlook to 2027, but to examine this it is necessary to understand key aspects of the global context. In particular, we will need to look at what might be a scientifically realistic emissions budget for the UK and to do this it is necessary to consider the global situation on climate change.

But for a true appreciation of the challenge of climate change it is necessary to understand the intertwined problem of peak oil, so this is where we will begin.

"Slowing and reversing these threats (the impacts of climate change) is the defining challenge of our age."
– Ban Ki-Moon, Secretary General of the United Nations, November 2007

Peak oil

Introduction to peak oil

It is a fact well-established by experience that the rate of oil production from a typical oilfield increases to a maximum point and then gradually declines. This is primarily a result of the geology of oil extraction and the finite amount of oil in any given field. This point of maximum flow is known as the production *peak*.

Because the same is true of the total oil production from a collection of oilfields the peaking concept is also applied to regions, to countries and to the entire world. This *global production peak* is what is generally referred to as *peak oil*.

The term *peak oil* is also commonly used as shorthand for energy resource depletion more generally, and the challenges associated with this. Due to our society's extraordinary dependence on oil, and the lack of comparable substitutes, oil is the main focus, but other non-renewable fuels such as natural gas, coal and uranium all face depletion issues to varying degrees of urgency.

A note on reserves and flow rates

It is often said that peak oil is the point at which half the oil has been extracted, but this can be a little misleading. While it is true that traditional oilfields would tend to hit their production peak when around half the oil had been extracted, nowadays, with developing extraction technologies and unconventional sources of fuel this has become a less useful rule of thumb.

For example, there are huge deposits of 'oil shale' in Colorado. This is essentially rock containing kerogen, which under the right circumstances – and with vast inputs of energy – can be heated to around 370°C to release oil. As Randy Udall, Director of the Community Office for Resource Efficiency put it,

> *"Shell has spent $200 million dollars to produce 1,700 barrels of shale oil in the last decade. At that rate of production the shale oil that we have here in Colorado will last six million years. This is something that gives me great hope for the future."* [128]

His tongue was firmly in his cheek. As we will see in a moment, current global oil production

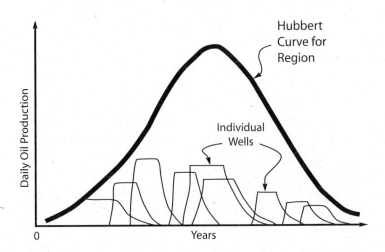

HUBBERT CURVE
Regional vs. Individual wells

*Figure 11: This type of oil production curve is known as a 'Hubbert curve', after the geologist M. King Hubbert. **Image courtesy of ASPO Italia.***

is around *74 million* barrels a *day* (b/d). Vast reserves of a poor energy source that we can only produce at a trickle are essentially irrelevant. As Udall says,

> "*Suppose you owned $100 million dollars, but the bank would only allow you to withdraw $10,000 each year. You would be rich . . . sort of.*" [129]

The constraint on production rates from Colorado oil shale, as with the unconventional oils in Canada and Venezuela, is not the depletion of reserves, but rather factors such as the environmental devastation and carbon emissions caused by their extraction, the amount of water needed and, ultimately, the low levels of useful energy they actually contain.[130]

Always bear in mind that depleting reserves are just one factor that can limit production and it is this rate of production (or *flow*) that determines how much energy is available to society at any given time – with peak oil it's not the size of the tank which is fundamental, but the size of the tap.

Where we are today

The very fact that we have started using these dirty and difficult-to-extract unconventional oils gives us a clue that the limits of flow rates from conventional oil are in sight. Figure 12

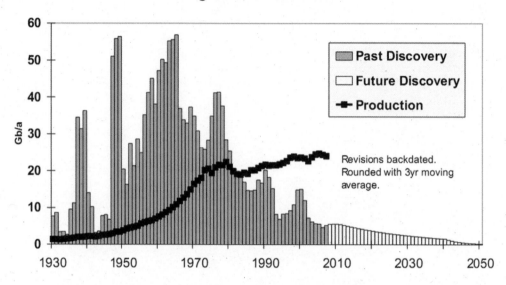

Figure 12: The Growing Gap.[131] *Gigabarrels/annum = Billion barrels/year.*
Source: ASPO Ireland Newsletter No. 96, December 2008.

shows that global *discovery* of conventional oil peaked in 1965, and that it is now over 25 years since we last discovered as much as we produced. Since we must find oil before we can produce (extract) it, it has been clear for some time that sooner or later global production will drop in line with discoveries.

Our next graph, Figure 13 (which runs to September 2008) highlights that global production of conventional oil broadly levelled off from mid-2005 to mid-2008, despite the incentive to maximise production caused by the massive increase in oil prices (from a $13 average per barrel in 1998 to over $140 in July 2008). Many new oil wells have begun

producing in this time, so this means that the new production has only just managed to offset the accelerating declines in production from existing fields. These production losses through depletion are only going to increase.[132]

Our third graph (Figure 14 overleaf) shows global 'total liquids' production, which adds other fuels such as the unconventional oils and biofuels to the total. This graph also runs two months further than the previous one, to November 2008. We can see here that even with all our efforts, such as turning around 30% of the US corn crop into biofuels, the world supply has been increasing slightly, but at a much reduced rate. Reaching 90 million barrels per

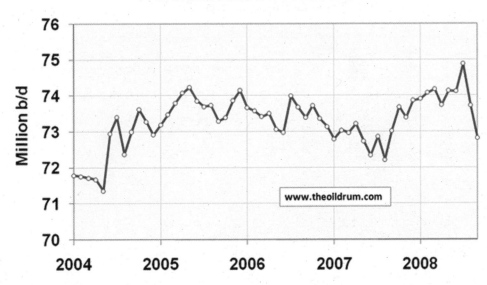

Figure 13: World Crude Oil Production.
Source: *Oilwatch Monthly by Rembrandt Koppelaar, December 2008, http://tinyurl.com/7mko6y*

day (an extraordinary flow rate, when you stop to think of it) looks like quite a struggle as more and more existing oil fields pass their peak.

There is heated debate over whether this levelling off in global oil supply is being caused by geological or geopolitical factors, but this is really beside the point. It is clear that both are factors – geopolitical factors in unstable regions do cause genuine disruption, but if there were no geological limits to production in more stable areas, the world would not be dependent on these more challenging supplies in the first place. The actual production figures we have just examined are the bottom line.[133]

Future oil demand

The problem is that mainstream projections of oil *demand* look like those given at the foot of this page.

It is this widening gap between supply and demand which has led to the huge increases in the oil price, and which is also leading to

Mainstream projections of oil demand.[133a]	International Energy Agency (IEA)	2015: 97 mb/d	2030: 106 mb/d
	US Dept. Of Energy (US DOE)	2015: 96 mb/d	2030: 113 mb/d

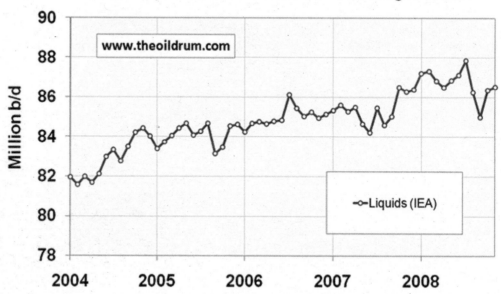

Figure 14 : World Liquid Fuel Production.
Source: *Oilwatch Monthly by Rembrandt Koppelaar, December 2008, http://tinyurl.com/7mko6y*

increasingly shrill voices warning that we have a serious problem, including from within the oil industry and other unexpected sources.

"Numbers like 120 million barrels per day will never be reached, never." – Christophe de Margerie, then Head of Exploration, Total. April 2006 [134]

"100m barrels [per day] . . . is now in my view an optimistic case. It is not my view: it is the industry view, or the view of those who like to speak clearly, honestly, and not . . . just try to please people." – Christophe de Margerie, now CEO, Total. October 2007 [135]

"I don't think we are going to see the supply going over 100 million barrels a day." – James Mulva, CEO, ConocoPhillips. November 2007 [136]

"There is no doubt demand for oil is outpacing supply at a rapid pace, and has been for some time now." – Rick Wagoner, Chairman and CEO, General Motors. January 2008 [137]

"Oil prices are going up because the demand for oil outstrips the supply for oil. . . . Oil is going up because we use too much oil, and the capacity to replace reserves is dwindling." – George W. Bush, President, United States of America, November 2007 [138]

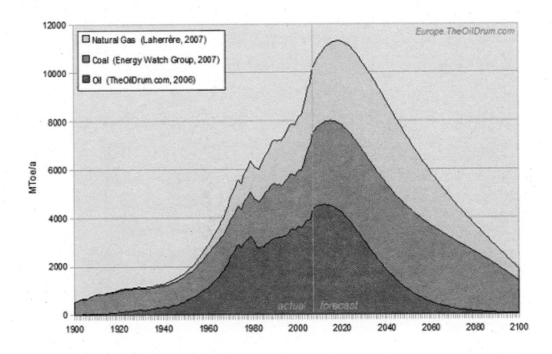

Figure 15: Global fossil-fuel production. In order to meaningfully compare the energy contained in the different fuels, this graph is quantified in Million Ton Oil Equivalents per annum (MToe/a) – a million barrels of oil per day is approximately 53 MToe per annum.[140] Source: www.theoildrum.com.

"One of the problems we've got now obviously is that there is not a lot of excess capacity worldwide." – Dick Cheney, Vice-President, United States of America, March 2008 [139]

Future fossil fuel prospects

Bear in mind that unlike the earlier graphs, Figure 15 is a projection, but it is a well-researched one, and clearly outlines the broad shape of the future for fossil fuels as depletion takes an ever-stronger hold. The evidence points to an inexorable decline, with the available flow of oil halving over the next few decades and other fossil-fuel energy sources unable to make up the shortfall. The decline will be even steeper if we limit our fossil-fuel consumption in the light of climate change.

Energy Return On Energy Invested

Energy Return On Energy Invested (EROEI or EROI) is an important concept when considering energy sources. Producing energy requires energy (e.g. the energy invested in building and installing solar panels), so the ratio between the energy you put in and the energy return you get out is crucial. EROEI tells you about the 'energy profit'.

The first oil discoveries, which flowed out of the ground under their own pressure, had an estimated EROEI of over 100:1 – they contained highly concentrated energy and took little energy to extract. Solar panels currently come in between 2:1 and 10:1, whereas some kinds of ethanol production actually drop below 1:1. In other words it takes more energy to produce the ethanol than is actually contained in the ethanol produced.[141]

The concept of EROEI, or *net energy*, becomes especially relevant when we consider the difference between the oil we extract today and those early oil discoveries. Naturally, we tend to extract the high quality easy-to-reach oil first,

so now we find that both sides of the EROEI ratio are worsening. We are extracting lower-quality oil that takes more energy to refine into useful forms, and we are also using more energy to extract it since it may be located, for example, a mile or two beneath the ocean.

Figure 16 illustrates this visually. The principle behind the graph applies to non-renewable energy resources generally, but let's consider oil. The line at the top of the black area represents the energy yield over time, as in Figure 15. But when we look within that we see that direct energy costs like drilling (**D**), indirect energy costs like refining (**C**), and environmental energy costs like cleaning up spills (**B**) are using up more and more of the energy produced, at the same time as the total energy yield is declining.

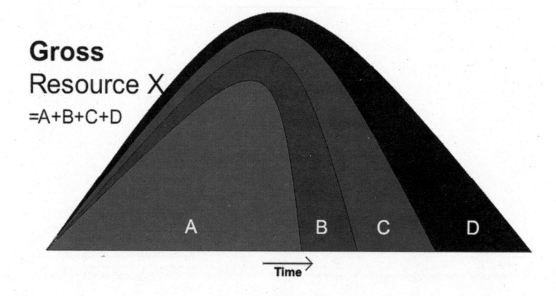

Figure 16: EROEI illustrated.[142] *Source: Nathan Hagens and Kenneth Mulder.*

It is only section (A) that represents the net energy available from the process, once we have extracted the oil and cleaned up after ourselves – *Section A represents the energy we can actually use to power society.* As we can see, the peak in net energy from oil comes before the production peak, and the decline in net energy is considerably faster than the decline in production of the resource itself.[143]

Oil prices

Figure 17 below shows yearly average oil prices from 1900-2007, and as we can see oil prices remained under $10/barrel until the oil crises of the 1970s (the graph also shows 'real' prices in 2007 dollars, adjusted for inflation, but we will stick to discussing the nominal price, which is the price at which oil was listed and sold at the time).[144]

The dramatic price increases at this time were due to political issues which caused major oil-exporting countries to reduce their oil production (as well as refusing to ship any oil to certain nations), and were seriously damaging to the economies of countries dependent on this oil. Nonetheless, oil production was able to recover after these crises passed and as we can see below, barrel prices fell back into the $15-$30 range until 2003 (prices were briefly under $10 as recently as February 1999, but this graph displays prices averaged over a year).[145]

The difference today is the underlying trend of supply and demand which we have just examined. The price surge shown at the right of our graph is not rooted in a temporary supply disruption that we can expect to be resolved in the near future, but rather in fears of the inexorable reality of geological depletion.

It appears that there is little or no remaining untapped supply to address either the predicted increases in oil demand (p.119), or

"According to normal economic theory, and the history of oil, rising prices have two major effects, they reduce demand and they induce oil supplies. Not this time."
– Fatih Birol, chief economist at the International Energy Agency, 'Oil Price Rise Fails to Open Tap', *New York Times*, April 9th 2008

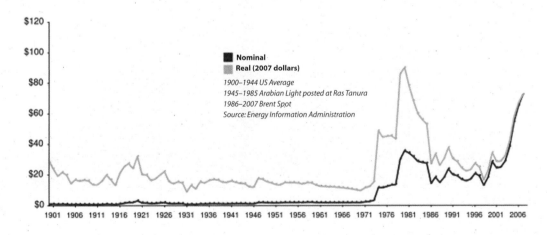

Figure 17: Historical Oil Prices (annual average) – 1900-2007.[146]

any disruptions to oil supply (whether rooted in geopolitics, geology, weather, engineering or any other cause). Consequently, when either of these come to pass, the price has to rise immediately in order to price people out of the market ('destroy demand') and bring supply and demand back into balance.

This means that the underlying price trend is likely to remain upward, but it also makes markets (and thus oil prices) very jumpy, with any information on higher-than-expected demand or lower-than-expected supply sending prices soaring, and any news in the opposite direction setting them plummeting.[147]

We saw this in 2008, with prices reaching over $140 a barrel in July (representing a 100% price increase in just one year), before falling to around $40 a barrel by December 2008, under pressure from the global economic downturn.

The only thing that can bring the price trend back down for a significant length of time is a sustained reduction in demand, whether caused by conditions like the present global economic hardship, or by a somewhat kinder transition to a less oil-dependent way of life.

Our graph above (Figure 17) ends with the 2007 average price of $72/barrel. Despite the low price at the end of the year, the 2008 average was substantially higher than that, yet the key point to remember is that it is the availability of useful energy and how sensibly we use it that will determine the future of our society. While the price may be the focus of all the media's attention, it is really a secondary factor.

The implications of peak oil

A 2005 report commissioned by the US Department of Energy concluded that,

"the peaking of world oil production presents the US and the world with an unprecedented risk management problem,"

and that without timely mitigation the economic, social and political impacts will be

"abrupt, revolutionary and not temporary".[148]

The reasons for this are detailed and complex, but ultimately it comes down to this – energy is the ability to do work of any kind, and oil is our most useful and most heavily used source of net energy. The implications of both increasing (and increasingly volatile) prices and actual oil supply shortages will be profound.

The sheer usefulness of oil can perhaps best be summed up with a stunning statistic. A 40-litre fill-up of petrol represents the energy equivalent of *four years* of manual labour by a person (as peak oil educator Richard Heinberg says, compare the effortlessness of driving fuelled by oil with the amount of muscle power it takes to push a car just to the side of the road). Yet in the UK we currently pay only around £45 for that amount of energy (and in the US they pay less than half that) – we would be hard-pressed to find someone willing to work for us for four years for that sum! [149]

Each 42-gallon barrel of oil yields around 20 gallons of petrol. We have seen that the world currently produces around 87 million barrels a day, so roughly speaking this works out at the energy equivalent of over 240 billion person-days of work contained in the world's daily petrol supply (quite apart from the diesel, jet fuel, heating oil etc. that we also produce from that oil). *Our current global petrol supply can do approximately 35 times as much physical work as every person on the planet put together.*

We take this available energy for granted much of the time in our everyday lives, but it is as though we had dozens of 'energy slaves' working for us day and night. It has been calculated that this energy input from oil allows the UK economy to be between 70 and 100 times more productive than would be possible on human muscle power alone.[150]

And in addition to being an abundant, reliable, cheap, super-concentrated form of energy, oil is also a liquid, making it far easier to transport, store and use than solid fuels. There are relatively few options for replacement liquid fuels, and since our vehicles and infrastructure are designed for oil, it would require technical innovation, a large investment of energy and other resources and a timeframe of at least twenty years to create an alternative system.[151]

This incredible energy source fuelled the rapid developments of the 20th century, whether in technology, industry, food yield or transport, and is also the source of the plastics and many synthetic materials that are everywhere around us. Ninety-five percent of all goods in shops involve the use of oil, and ninety-five percent of the UK's food is now oil-dependent. Just to farm a single cow and deliver it to market requires six barrels of oil, enough to drive a car from New York to Los Angeles.[152]

As oil becomes more expensive and less available it affects the price and availability of the products and services throughout the economy that are dependent on it, as well as the jobs tied into these products and services. And since oil features in the supply chain of almost every company, the health of the national and global economy is also threatened as they all find their costs increasing within an economy whose total productive capacity is decreasing.

In other words, the growth of our economy is dependent on a growing net energy supply, and for the first time in centuries it is unlikely to have it. A real cause for concern is that our economic system as currently designed fails without continued growth, leading to bankruptcies, defaults on loans and mortgages, mass unemployment, homelessness and a myriad of other unpleasant consequences.

At this point, however, it seems appropriate to remember the insight of the science fiction writer William Gibson that:

"The future is already here, it's just unevenly distributed."

Peak oil is no longer simply a future problem. While richer countries have been able to pay the increasing prices demanded for oil globally, those with less money have been struggling to afford the supplies they rely on.

In early 2007, a U.N. report found that:

"Recent oil price increases have had devastating effects on many of the world's

"Fuel is our economic lifeblood. The price of oil can be the difference between recession and recovery. The Western world is import-dependent".
– Tony Blair, Speech at the George Bush Senior Presidential Library, April 7th 2002

poor countries, some of which now spend as much as six times as much on fuel as they do on health. Others spend twice the money on fuel as they do on poverty alleviation. And in still others, the foreign exchange drain from higher oil prices is five times the gain from recent debt relief.

Of the world's 50 poorest countries, 38 are net importers of oil and 25 import all of their oil requirements." [153]

When the report was published the oil price stood at $60/barrel, but over the next 18 months it proceeded up to over $140 a barrel. If the effects on the poorest countries were "devastating" at $60/barrel, they became yet more so.

The possible future America was warned of in that 2005 Department of Energy report, with its abrupt and revolutionary economic, social and political impacts, has already been unfolding elsewhere. Outright energy shortages and deadly fuel riots have been seen across the world, and the peak oil predicament underlying them is only worsening as time goes by. [154]

In Part Five (p.154) we will examine the direct implications of peak oil for the UK, and our Government's position on the matter, but let us now consider how the peak oil issue interacts with the challenge of climate change.

Chapter 16

Peak oil and climate change – the interplay

When we consider peak oil and climate change together, an interesting question arises. Is peak oil a useful factor, limiting the availability of the fossil fuels which are destabilising our climate, or does it make the climate challenge even greater by presenting us with limits to our energy availability and economic viability just as we are faced with our greatest challenge? There is no straightforward answer to this question, but we can make a few important observations.

If we were trying to take the right decision from a pure climate-change perspective, we might be tempted to simply ban the extraction of any more fossil fuels from the ground today, but this would be at the cost of unacceptable suffering due to the sudden loss of the lifeblood of our fossil-fuel-based societies (our infrastructure for food supply, transportation, heating, electricity and so forth would rapidly fail catastrophically).

So there is clearly a tension between addressing climate change and addressing peak oil. The earlier we reach fossil-fuel supply limits – whether geological or voluntary – the better for climate change, but the more painful the 'peak oil' adaptation problems, and the higher the oil price.[155]

As the oil price rises, more and more countries (and ultimately individuals) are priced out of the market, leaving only those with enough money able to get the oil their lifestyles demand. Economists call this *demand destruction*, and it is the mechanism the market uses to close the widening gap between supply and demand.

Unfortunately, markets do not distinguish between more and less essential uses of oil. If we in the UK are willing and able to pay more to run our cars than people elsewhere are able to pay to heat their homes or power their hospitals then the limited supply of oil will flow here, regardless of the suffering caused to those whose demand is 'destroyed', as discussed in the previous chapter.

The international oil price, then, is effectively a rough measure of how much of this painful demand destruction is going on. The more limited the supply the more demand destruction is necessary and the higher the price goes.

On reading the peak oil literature it is clear that the concerns revolve around depletion and shortage. However, it is also clear that there is more than enough carbon in the fossil fuels we have discovered to bring about catastrophic changes in our climate and environment. There is the clear necessity of leaving existing fossil fuels in the ground.

It is rarely considered that this turns the scarcity paradigm on its head, as we could find ourselves in a situation in which unexploited fossil fuel reserves are regarded as valuable but dangerous resources which need careful supervision or even guarding by governments to prevent their extraction. We are used to considering resource depletion in terms of having too little of a good thing. We may need to start thinking that there is effectively too much of a bad thing.

The supply side dilemma

The question of whether we should leave some of the available fossil fuels in the ground, then, becomes a question of whether the effects of increased oil demand destruction are more or less desirable than the effects of increased emissions and the resultant climate change.

This is what I call the 'supply side dilemma', and it leaves us attempting to choose the lesser of two evils. Not much of a choice some might say, and it is easy to see why there are passionate advocates on both sides of the debate.

Thankfully though, there are things we can do to ameliorate both climate change and peak oil simultaneously. If we begin to wean our communities off their oil addiction voluntarily, then we reduce demand, and thus reduce the need for the more painful varieties of demand destruction. By reducing the desperation for increased fossil fuel supplies we also make it easier to consider the necessary step of leaving some of it where it is as a response to climate change.[156]

The more imaginative and creative ways we can find to reduce demand, the less difficult the supply side dilemma becomes.

The substitution problem

One issue that is often raised when considering the interplay of peak oil and climate change is the 'substitution problem' – that, as the availability of relatively clean (low emissions) and useful (high EROEI) liquid fuels declines, we are liable to turn to less clean and useful fuels in an attempt to maintain the convenient energy inputs on which our way of life depends. The development of the Alberta tar sands is an example of this – they produce heavy emissions for a very low EROEI, and, if sufficient easily extractable high quality oil were available, we would certainly be leaving them well alone (see p.121 for a discussion of EROEI).

Some fear that exploiting such dirty alternative fuels may cause our total carbon emissions to increase, even as emissions from conventional oil and natural gas production decline in line with depletion. However, it is not widely recognised that on the timescales we are considering (looking out to 2027), the scale of probable oil production declines means that *even the maximum feasible expansion of alternative liquid fuels could not maintain current emission growth rates.*[157]

In other words, although fossil-fuel depletion will present many challenges that compound the climate emergency, it will at least slow the growth in emissions from liquid fuel use. As we will see, these involuntary limits on fuel emissions are not in themselves anywhere near sufficient to address the climate problem, but they do make apparent one critical consequence.

Peak oil will reduce our emissions while also leading us to use fuel sources which generate less useful energy per unit of emissions. This means that the total useful energy available to us is set to decrease rapidly, and this end to abundant cheap energy signifies the end of our current energy-intensive way of life.

Since energy represents our ability to do any kind of work this presents serious challenges on many levels, and it is our possible responses to these challenges – and the opportunities that come with them – that are the focus of Parts One, Two and Three of this book.

(Right) Rob Hopkins discussing the challenges and opportunities of peak oil and climate change.

Chapter 17

Climate change explained

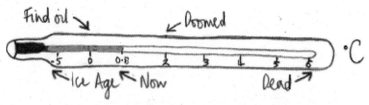

Figure 18 : The Climateometer. Used by kind permission of www.ageofstupid.net.

We are all familiar with the concept of climate change, and the need for reduced carbon emissions, but really getting a handle on the scale of the problem can be difficult, thanks to all the confusing terminology.

This section of the book will explain the meaning of terms like Global Warming Potential (GWP) and the different meanings of 'CO₂ equivalent', before going on to examine both the present position and the future risks. It will at times refer to the work of the Intergovernmental Panel on Climate Change (IPCC), the body established jointly by the United Nations and the World Meteorological Organisation in 1988 to assess the available scientific evidence.[158]

In order to fully understand the relationship between greenhouse gas emissions and global temperature increase then, we first need to consider the concept of *radiative forcing*.

The Earth is continually receiving energy from the Sun, and continually losing energy into space (as space is much cooler than

the Earth). Radiative forcing is simply *the difference (measured in watts per square metre) between the amount of energy received and the amount of energy re-radiated back into space.* In other words, it is the rate at which the planet's surface is either warming or cooling.[159]

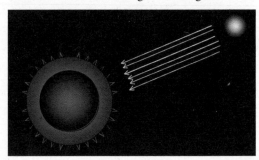

Figure 19: Radiative forcing.
Used by kind permission of David Wasdell

If the planet were losing energy at the same rate it was gaining it then the radiative forcing would be zero and the temperature would remain stable at its current level – this state is called *thermal equilibrium.* Since a hotter planet loses more energy into space, the natural system

Business As Usual (BAU)

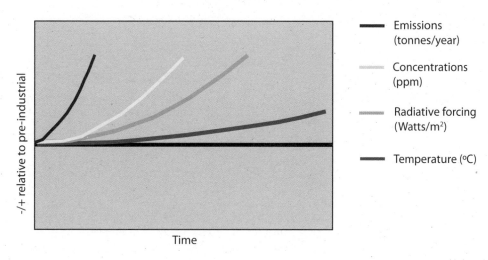

Figure 20: Climate change under Business As Usual.

tends to move towards thermal equilibrium.

However, rising greenhouse gas concentrations (measured in parts per million – ppm[160]) in the atmosphere act like an insulating blanket, reducing the rate at which energy can escape into space, and so affecting radiative forcing, which in turn affects the temperature. The rough illustrative graphs (Figures 20 and 21) give an idea of these relationships and show the time delay between changes in emissions rates (up or down) and temperature changes.[161]

Figure 21 shows that if we can bring anthropogenic (human-caused) emissions back down we can stabilise greenhouse gas concentrations and bring radiative forcing back towards equilibrium, but at a higher temperature.[162]

So, emissions contribute to greenhouse gas concentrations which in turn contribute to radiative forcing, but it is radiative forcing that determines the rate of change in temperature. Armed with this understanding, the terms below become clearer:

Global warming potential (GWP) is an estimate of how much a given greenhouse gas contributes to Earth's radiative forcing. Carbon dioxide (CO_2) has a GWP of 1, by definition, so a gas with a GWP of 50 would increase radiative forcing by 50 times as much as the same amount (mass) of CO_2. A GWP value is defined over a specific time interval, so the length of this time interval must be stated to make the value meaningful (most researchers and regulators use 100 years).

For example, methane has a GWP of 72 over 20 years, but a lower GWP of 25 over 100 years. This is because it is very potent in the short-

Emissions reductions

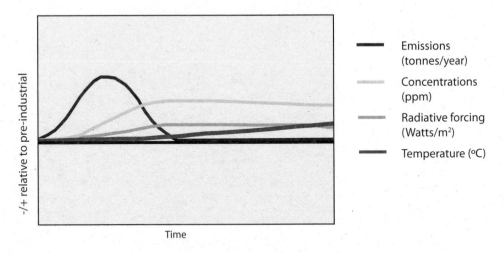

Figure 21: Climate change with emissions reductions.

term but then breaks down to CO_2 and water in the atmosphere, meaning that the longer the period you consider it over, the more similar its effect is to that of CO_2 alone.[163]

Equivalent carbon dioxide (CO_2e) is an estimate of the concentration of CO_2 (in ppm) that would cause a given level of radiative forcing.[164]

For example, the IPCC's latest report in 2007 considered the effects of the main greenhouse gases currently present in our atmosphere and calculated a CO_2e for these of around 455ppm (and rising). This means that (over a defined period) the radiative forcing effect of these gases at current concentrations is roughly equal to the effect a 455ppm concentration of CO_2 alone would cause. This particular CO_2e calculation takes into account the six major greenhouse gases considered under the Kyoto Protocol, and so may be labelled **CO_2e(Kyoto)**.[165]

However, the radiative forcing line in Figures 20 and 21 represents the total radiative forcing of the planet. This is the important figure – the one that determines the rate of change in Earth's temperature – and as well as the Kyoto gases it is also affected by other factors such as the effects of sulphate aerosols, ozone and cloud formations. Figure 22 quantifies the effect of each of these factors, and we can see that a number of them (those coloured blue) are actually *negative forcings*, which act to reduce the total radiative forcing. Because of these, the equivalent CO_2 for *all forcings combined* – **CO_2e(Total)** – is, thankfully, lower than CO_2e(Kyoto). The IPCC's latest figures give CO_2e(Total) as roughly 375ppm.[166]

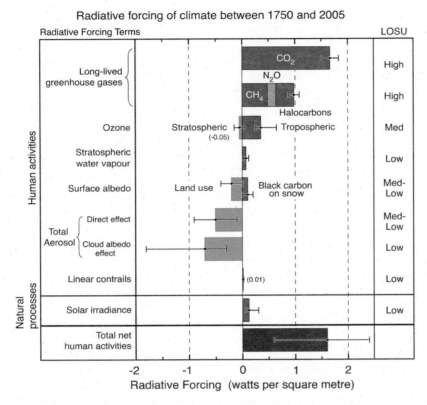

FAQ 2.1, Figure 2. *Summary of the principal components of the radiative forcing of climate change. All these radiative forcings result from one or more factors that affect climate and are associated with human activities or natural processes as discussed in the text. The values represent the forcings in 2005 relative to the start of the industrial era (about 1750). Human activities cause significant changes in long-lived gases, ozone, water vapour, surface albedo, aerosols and contrails. The only increase in natural forcing of any significance between 1750 and 2005 occurred in solar irradiance. Positive forcings lead to warming of climate and negative forcings lead to a cooling. The thin black line attached to each coloured bar represents the range of uncertainty for the respective value. (Figure adapted from Figure 2.20 of this report.)*

Figure 22: Factors affecting planetary warming/cooling.
Source: http://www.ipcc.ch/pdf/assessment-report/ar4/wg1/ar4-wg1-spm.pdf – Figure 2, p.4.

When we hear scientific debates between stabilisation scenarios of, say, 350ppm, 450ppm or 550ppm it is CO₂e(Total) which is under discussion. So this 375ppm is the key number, but it has a far wider margin of error than the others. This is because it is relatively easy to measure the atmospheric concentrations of greenhouse gases, and the GWP of those gases, but considerably more difficult to account for all the effects that contribute to the ultimate CO₂e(Total) radiative forcing over a given period. The column labelled LOSU in

Figure 22 stands for the 'Level Of Scientific Understanding' of the various forcings, and as we can see it is not universally high.

Radiative forcing is the fundamental issue, but it is easy to see why most discussions revolve only around emissions – not only are CO_2 emissions much the largest way in which humanity is changing the planet's radiative forcing, but they are also easier to understand conceptually and easier to quantify than radiative forcing.

According to the IPCC, atmospheric CO_2 concentrations were 379ppm in 2005, which coincidentally happens to be close to our best estimate of 375ppm CO_2e(Total). Unfortunately this coincidence also creates a good deal of confusion, as it is not always clear which measure an author is referring to – scientists often assume that this is obvious to their audience, and many others do not themselves fully understand the distinctions between CO_2, CO_2e(Kyoto) and CO_2e(Total).[167]

The other source of confusion is that all of the numbers we have discussed are based on evolving science, and many can only be given approximately. For example, these are the IPCC's given figures for the GWP of methane over 100 years, taken from their last three reports:

- 1995 – 2nd Assessment Report (SAR): *Methane 100 year GWP = 21*

- 2001 – 3rd Assessment Report (TAR): *Methane 100 year GWP = 23*

- 2007 – 4th Assessment Report (AR4): *Methane 100 year GWP = 25*

These changes are entirely appropriate – the values should become more accurate over time as new measurement methods or changes in scientific understanding develop – but it makes it important to check where any figures are sourced from.[168]

If you followed everything in the past few pages, you are now well-equipped to consider the scientific discussion of climate change. You may find you understand it better than some of those who write and speak about it!

Where we are today

So let's take stock. Below are the latest IPCC figures, which define the situation as it was in 2005:

- CO_2 = 379ppm (error range: minimal)

- CO_2e(Kyoto) = 455ppm (error range: 433-477ppm)

- CO_2e(Total) = 375ppm (error range: 311-435 ppm) [169]

Annual CO_2 emissions increased by about 80% just between 1970 and 2004. Atmospheric CO_2 concentrations are currently rising by between 1.5 and 3 ppm each year and were at roughly 385ppm in mid-2008. It is worth noting that the pre-industrial concentration of CO_2 in our atmosphere was 278ppm and did not vary by more than 7ppm between the years 1000 and 1800 C.E.[170]

We are now in a position to examine the likely consequences of the changes we are causing to our global climate system. For the reasons outlined above we will largely stick to discussions of CO_2, only bringing in CO_2e of any kind when necessary, and labelling it clearly when this is done.

Chapter 18

Climate change – the IPCC

The Intergovernmental Panel on Climate Change (IPCC) in many ways represents the scientific orthodoxy on climate change. It was established twenty years ago and today involves scientific experts from more than 130 countries – its most recent report (2007's Fourth Assessment Report, or 'AR4') had over 800 authors, and was peer-reviewed by another 1,000+ experts. We have used the IPCC's highly reliable data in establishing where we stand today, so their work is the obvious starting point for examining the likely future impacts.[171]

IPCC scenarios and peak oil

As throughout this book, when considering the IPCC's work we will do so informed by an awareness of peak oil. In 2000 the IPCC defined a number of scenarios for future man-made CO_2 emissions, which were theoretically regarded as equally likely. The less ecologically friendly scenario families are called A1 and A2, and the more ecologically friendly B1 and B2. According to the authors:

> "These scenarios cover a wider range of energy structures than the (IPCC's 1992) scenarios. This reflects uncertainties about future fossil resources and technological change. The scenarios cover virtually all the possible directions of change, from high shares (use) of fossil fuels, oil and gas or coal, to high shares (use) of non-fossils." [172]

However, despite these claims none of these scenarios takes a realistic view of supply limits on fossil fuels, and when we consider the growing body of evidence on depletion, it becomes clear that resource limits mean we cannot extract fossil fuels fast enough to generate the carbon emissions projected by most of these scenarios.

While our global fossil-fuel usage generated emissions of almost 8 Gigatonnes (1 Gt = 1 billion tonnes) of carbon in 2004, the IPCC's A1F and A2 scenarios project emissions in 2100 from fossil fuels alone of 30.3 GtC/yr and 28.9 GtC/yr respectively – more than triple current fossil fuel usage! Even taking into account the most abundant estimates of remaining fossil fuel reserves, it appears a physical impossibility to steadily increase energy emissions to that level, and for that we might be thankful. As graph (b) in Figure 23 shows, if scenario A2 were possible it would lead to atmospheric CO_2 concentrations of more than 850ppm by 2100, which would increase the rate of global

"If there's no action before 2012, that's too late. What we do in the next two to three years will determine our future. This is the defining moment."
– Rajendra Pachauri, the scientist and economist who heads the IPCC, quoted in 'Alarming UN report on climate change too rosy, many say', *International Herald Tribune*, November 18th 2007.

"If you look at the scientific knowledge, things do seem to be getting progressively worse, so you'd better start with the interventions even earlier. Now."
– Rajendra Pachauri, in a later interview the same day, quoted in 'Alarming UN report on climate change too rosy, many say', *International Herald Tribune*, November 18th 2007

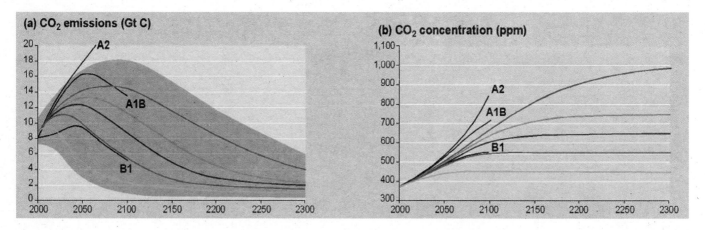

Figure 23 (above):
CO2 emissions and
concentrations under
different IPCC scenarios.
Source: IPCC http://www.ipcc.
ch/graphics/graphics/2001syr/
large/02.18.jpg

temperature change far beyond the adaptive capabilities of much life on Earth.[173]

In an April 2007 paper entitled 'Implications of peak oil for atmospheric CO_2 and climate' Drs Pushker Kharecha and James Hansen of NASA confirm that scenario A2 is unrealistic. Figure 24 illustrates their calculation that the peak in CO_2 concentrations reaches 'only' 580ppm even if all available fossil fuels are burned, once consideration of peak oil and the Energy Information Agency's (probably overstated) figures on fossil fuel reserves limits are incorporated.[174]

However, we must be careful here. Kharecha and Hansen note that:

"The contribution of unconventional fossil fuels to CO_2 emissions is negligible to date. We do not include unconventional fossil fuels in the scenarios that we illustrate, because we are interested primarily in scenarios that cap atmospheric CO_2 at 450 ppm or less. However, it should be borne in mind that unconventional fossil fuels could contribute huge amounts of atmospheric CO_2, if the world should follow an unconstrained 'business-as-usual' scenario of fossil fuel use." [175]

Such unconventional fossil fuels include oil shales, tar sands and methane hydrates, and higher global emissions scenarios are also possible due to non-fuel-based scenarios

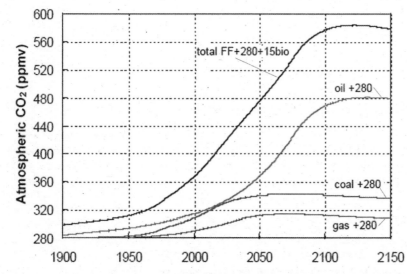

Figure 24: CO2 concentrations if all available fossil fuels are burned.
Source: http://www.theoildrum.com/node/2559

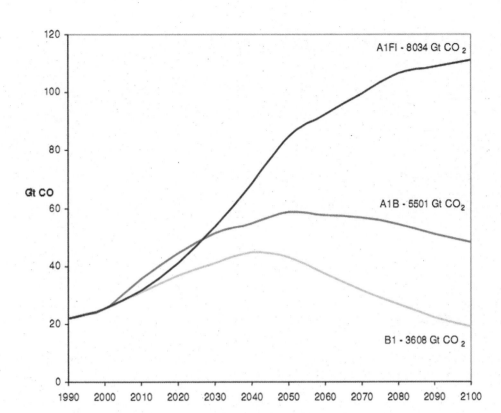

Figure 25: Annual emissions under different IPCC scenarios.[178]
Source: Garnaut Review Interim Report, http://tinyurl.com/2q9wnv , p.15.

such as the collapse of the Amazon rainforest through drought and fire, but in such scenarios climate change would likely be entirely beyond human control. As with Kharecha and Hansen our focus is on identifying scenarios that might avoid such possibilities and, quite apart from the climatic impacts, our understanding of the poor EROEI of unconventional fossil fuels also gives us good reason to consider the large-scale exploitation of unconventional fossil fuels both an unwise and an unlikely scenario.[176]

As such, from now on we will focus on the B1 emissions scenario, which is the IPCC's lowest projection of future emissions, but which still projects total emissions reaching almost twice 1990 levels by 2050, before returning to a little below 1990 levels by 2100 (the lowest line in Figure 25).[177]

We can see from graphs (b) and (c) in Figure 26 overleaf that even this lowest IPCC emissions scenario (B1) is projected to lead to CO_2 concentrations of 550ppm and over two degrees of warming by 2100, so let us examine

"We now have a choice between a future with a damaged world or a severely damaged world."
– Professor Martin Parry, co-chairman of the IPCC impacts working group, 'How climate change will affect the world', *The Guardian*, September 19th 2007

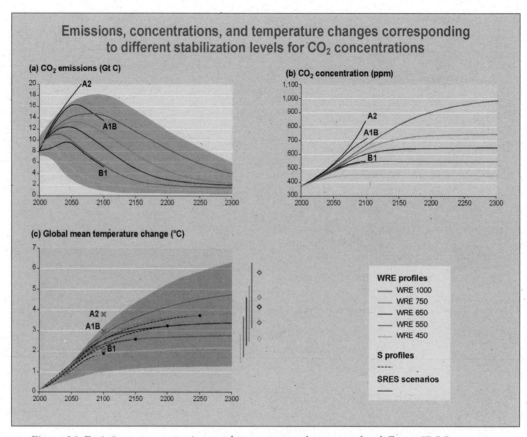

Figure 26: Emissions, concentrations and temperature changes under different IPCC scenarios.
Source: IPCC http://www.ipcc.ch/graphics/graphics/2001syr/large/02.18.jpg

the IPCC's findings on what impacts this level of climate change might have.

IPCC on the impacts of climate change

The IPCC reported in Sept 2007 that:

"If warming is not kept below 2⁰C, which will require the strongest of mitigation efforts, and currently looks very unlikely to be achieved, then substantial global impacts will occur, such as: species extinctions and millions of people at risk from drought, hunger and flooding, etc." [179]

The IPCC strictly define the boundaries of "very unlikely" – it means a likelihood of less than 10%.

Figure 27 shows why keeping warming below 2⁰C (above pre-industrial temperatures) is considered so unlikely. Remembering that

Table SPM.5: *Characteristics of post-TAR stabilization scenarios [Table TS 2, 3.10][a]*

Radiative forcing (W/m^2)	CO_2 concentration[c] (ppm)	CO_2-eq concentration[c] (ppm)	Global mean temperature increase above pre-industrial at equilibrium, using "best estimate" climate sensitivity[b], [c] (°C)	Peaking year for CO_2 emissions[d]	Change in global CO_2 emissions in 2050 (% of 2000 emissions)[d]
2.5-3.0	350-400	445-490	2.0-2.4	2000-2015	-85 to -50
3.0-3.5	400-440	490-535	2.4-2.8	2000-2020	-60 to -30
3.5-4.0	440-485	535-590	2.8-3.2	2010-2030	-30 to +5
4.0-5.0	485-570	590-710	3.2-4.0	2020-2060	+10 to +60
5.0-6.0	570-660	710-855	4.0-4.9	2050-2080	+25 to +85
6.0-7.5	660-790	855-1130	4.9-6.1	2060-2090	+90 to +140

a) The understanding of the climate system response to radiative forcing as well as feedbacks is assessed in detail in the AR4 WGI Report. Feedbacks between the carbon cycle and climate change affect the required mitigation for a particular stabilization level of atmospheric carbon dioxide concentration. These feedbacks are expected to increase the fraction of anthropogenic emissions that remains in the atmosphere as the climate system warms. Therefore, the emission reductions to meet a particular stabilization level reported in the mitigation studies assessed here might be underestimated.

b) The best estimate of climate sensitivity is 3°C [WG 1 SPM].

c) Note that global mean temperature at equilibrium is different from expected global mean temperature at the time of stabilization of GHG concentrations due to the inertia of the climate system. For the majority of scenarios assessed, stabilisation of GHG concentrations occurs between 2100 and 2150.

d) Ranges correspond to the 15th to 85th percentile of the post-TAR scenario distribution. CO_2 emissions are shown so multi-gas scenarios can be compared with CO_2-only scenarios.

Figure 27: A summary of the implications of different emissions scenarios.
Source: IPCC: http://www.ipcc.ch/pdf/assessment-report/ar4/wg3/ar4-wg3-spm.pdf, p.15

we have already reached atmospheric CO_2 concentrations of around 385ppm, we can see from the table that even if CO_2 concentrations remained at current levels the IPCC predict 2.0-2.4 degrees of ultimate warming. Keeping concentrations within these bounds would involve a peak in CO_2 emissions by 2015, with 50-85% reductions in global emissions by 2050 (relative to emissions rates in 2000).[180]

They conclude that with this level of warming Africa, the Asian megadeltas, small islands and the Arctic will be the areas worst affected, but that nearly all European regions are also anticipated to be negatively affected by some future impacts of climate change, including increased risk of inland flooding,

more frequent coastal flooding and increased erosion. With regard to Europe they state that:

"The great majority of organisms and ecosystems will have difficulty adapting to climate change." [181]

Just as we saw with peak oil, they also highlight that climate change can no longer be regarded as a future problem, stating that important effects of anthropogenic warming are already being felt. In 2007 IPCC Working Group II reported that observed effects of anthropogenic climate change can now be detected on every continent and are happening faster than predicted in the IPCC's earlier work. This is observed reality, not modelling:

"We are all used to talking about these impacts coming in the lifetimes of our children and grandchildren. Now we know that it's us."
– Professor Martin Parry, co-chairman of the IPCC impacts working group, 'How climate change will affect the world', *The Guardian*, September 19th 2007

"The effects of climate change are being felt now: temperatures are rising, icecaps and glaciers are melting and extreme weather events are becoming more frequent and more intense . . . above all, climate change is already being felt around the globe."
– 'Climate Change and International Security: Paper from the High Representative and the European Commission to the European Council' (2008)

"Inez Fung at the Berkeley Institute of the Environment says that for her research to be considered in the 2007 IPCC report, she had to complete it by 2004. 'There is an awful lag in the IPCC process,' she says, also noting that the special report on emission scenarios was published in 2000, and the data it contains were probably collected in 1998. 'The projections in the 2007 IPCC report [using the 2000 emission scenarios] are conservative, and that's scary.'"
– *Climate Code Red: The Case for a Sustainability Emergency*, David Spratt and Philip Sutton, Friends of the Earth Australia (2008)

"Observational evidence from all continents and most oceans shows that many natural systems are being affected by regional climate changes, particularly temperature increases." [182]

We can also note here that all but one of the UN's emergency appeals for humanitarian aid in 2007 were climate-related.[183]

Looking into the future, the IPCC see water supply, agriculture, human health and certain ecosystems (such as coral reefs) as being most heavily disrupted as our global climate shifts. And if we allow warming to continue:

"as global average temperature increase exceeds about 3.5⁰C, model projections suggest significant extinctions (40-70% of species assessed) around the globe." [184]

Clearly this is difficult news, but there are a number of reasons why the IPCC assessment of climatic risks and impacts is likely to be too conservative, and it is to these that we must now turn.[184a]

Problems with the IPCC approach
Slow response to new evidence

The most obviously problematic feature of the IPCC approach is its four-year process of drafting and review before final publication. This is arguably unavoidable for a body which aims to be the epitome of scientific rigour in the field but means that the final assessment will always lag behind the latest scientific evidence. This delay means that whether the balance of evidence is moving towards stronger or weaker climate impacts the IPCC modelling will be behind the times.

The balance of evidence

Unfortunately, over the last couple of years it has continually been the 'worst-case' scenarios from the models, or sometimes even worse, that have actually unfolded in reality.

The three graphs in Figure 28 show the actual measured changes in CO_2 concentrations, temperature change and sea level change since 1973, compared with the full range of IPCC scenarios (published 2000, shown as dashed lines and grey ranges). We can see that CO_2 concentrations have increased only slightly more than projected, but the impacts of that increase have been greater than the IPCC models predicted, with temperature change near the upper limit of the IPCC's range of possibilities, and sea level rise greatly exceeding even the worst-case scenarios from the modelled scenarios.

Sadly, this trend is continuing, even since the graphs in Figure 28 were drawn. In particular, the dramatic melting of sea ice in the Arctic summer of 2007 was far more severe than predicted by climate models.

This Arctic ice loss trend is shown in Figure 29, and is another which is exceeding the 'worst-case' predictions of the IPCC (in this case using the SRES A2 scenario). The smoothed satellite data represented by the bold black line runs to Sept 2007.

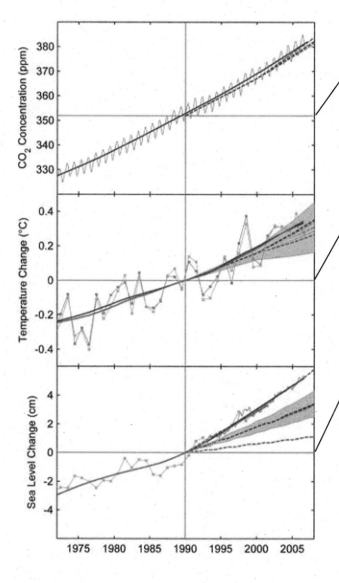

Figure 28: Actual observations compared with IPCC scenarios.[185]

CO2: *The oscillating solid line shows measurements from Mauna Loa, Hawaii up to Jan 2007. The bold line shows the trend.*

Temperature: *The oscillating solid lines show observed changes in annual global-mean land and ocean combined surface temperature from NASA GISS and the Hadley Centre up to 2006. Bold lines show their trends.*

Sea level: *The oscillating solid lines show observed changes based on tide and satellite data up to mid-2006. Bold lines show their trends.*

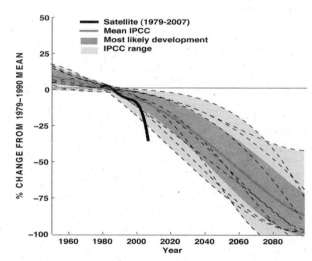

Figure 29 (right): Comparing actual Arctic sea ice loss with IPCC projections.
Source: www.carbonequity.info/images/seaice07.jpg

Climate feedbacks

It is clear that observed reality is exceeding the limits of the IPCC models, but the obvious question is why. What is lacking from their models? One key issue is that the most recent science they could incorporate is unable to account for the full range of feedbacks in the climate system, as they admit:

> *"The emission reductions to meet a particular stabilisation level reported in the mitigation studies assessed here might be underestimated due to missing carbon cycle feedbacks."* [186]

In this context, feedbacks are those responses of the planet to increased temperatures and greenhouse gas concentrations which themselves affect temperatures and greenhouse gas concentrations. Negative feedbacks (like the Earth losing more heat into space as it warms up) act to dampen radiative forcing. Positive feedbacks (like permafrost melting as the planet warms and releasing trapped methane reservoirs into the atmosphere) act to amplify it. Around 20 of these feedback mechanisms have been identified, and unfortunately the positive (amplifying) feedbacks overwhelm the negative (damping) feedbacks on the timescales we are concerned with. As Dr Hansen put it in 2007, "this allows the entire planet to be whipsawed between climate states." [187]

One particularly significant form of feedback concerns our planet's ability to absorb carbon. There are two main 'carbon sinks' – the oceans and the biosphere. Between them they have taken around half of human carbon emissions out of our atmosphere to date, but they are becoming less effective at doing so, for many reasons, including:

- more CO_2 in the oceans makes it more acidic, reducing its own ability to absorb further CO_2 as well as killing plankton that would itself absorb CO_2

- warmer water also absorbs less CO_2 and kills plankton

- plants become more vulnerable to drought, fire and disease as their local climate changes

- the stress of increasing temperature and CO_2 eventually causes plants to start releasing CO_2 rather than taking it up. [188]

A 2007 paper in the scientific journal of the US National Academy of Sciences estimated that the degradation of these carbon sinks accounts for about 18% of the increase in the atmospheric CO_2 growth rate since the year 2000, and confirmed that we are seeing "stronger-than-expected and sooner-than-expected climate forcing" which is likely to intensify. [189]

In what's becoming a familiar refrain, Ian Totterdell, a climate modeller at the Met Office Hadley Centre, said of the degrading oceanic carbon sink that:

> *"It's one of many feedbacks we didn't expect to kick in until some way into the 21st century."* [190]

The deeper problem though is that as each feedback mechanism works to raise concentrations and temperatures, those rising concentrations and temperatures feed into and amplify *all the other positive feedbacks.* This is called second order feedback, and it is this that may be changing our world faster than any of the IPCC projections foresaw.[191]

The IPCC also effectively ignores certain long-term feedbacks. The models that they use do attempt to incorporate the effects of water vapour, clouds and sea ice, but longer-term variables like ice-sheet area, sea level and vegetation distributions are assumed to be fixed in their climate sensitivity models. Drs Hansen and Sato argue that while this gives a fairly accurate model over the short timescales for which we have observations of anthropogenic climate change, over longer timescales the surface temperature is likely to be *twice as sensitive* to CO_2 concentration levels as the IPCC estimate.[192]

The IPCC admit that:

> "*The response of the climate system to anthropogenic forcing is expected to be more complex than simple cause and effect relationships would suggest; rather, it could exhibit chaotic behaviour with cascades of effects across the different scales and with the potential for abrupt and perhaps irreversible transitions.*"[193]

This is exactly what appears to be happening already, and so humanity must be very careful, because while all these feedback mechanisms and cascades of effects are kicking in, we only have direct control over our own emissions.

We have been pushing the boulder along and it may now be starting to roll downhill under its own momentum. This is what is meant by a climate tipping point.

And it is almost certain that there is some level of temperature rise at which the runaway feedbacks become more powerful than our ability to damp them back down. If we reach this point – where the boulder's momentum is too great for us to do anything to stop its headlong tumble – we are into *unstoppable climate change.*

Politically involved science

The Intergovernmental Panel on Climate Change (IPCC) is not, as many suppose, a group of scientists. It is rather, as its name suggests, a panel comprising representatives from about 140 governments. The panel decides whether an assessment is needed, and then engages scientists to conduct it (unpaid). Its role is to produce a summary of scientific understanding which, once commissioned and adopted, becomes accepted by the governments of the world (even the United States of America under George W. Bush signed off on it).

In order to ensure this acceptance, after the long process of scientific debate to reach a report which outlines what we know and where debate remains, the scientists then have to take their work line-by-line to the IPCC politicians to negotiate what gets included in the crucial 'Summary For Policymakers' (SPM).[194]

Professor Martin Parry, co-chairman of the IPCC impacts working group has said that:

> "*Governments don't like numbers, so some numbers were brushed out of it.*"[195]

"Two degrees has the potential to lead to three or four degrees because of carbon-cycle feedbacks."
– Professor Barry Brook, "No return' fears on climate change', *The Age*, June 12th 2008

"The policy-makers' summary, presented as the united words of the IPCC, has actually been watered down in subtle but vital ways by governmental agents before the public was allowed to see it."
– Peter Wadhams, ocean physicist and IPCC referee, 'Climate report was "watered down"', *New Scientist*, March 10th 2007

"Most of the climate change community, steered by Kyoto and IPCC, limit the scope of their consideration to the year 2100. By setting up the problem in this way, the calculation of a safe CO_2 emission goes up by about 40%, because it takes about a century for the climate to fully respond to rising CO_2. If CO_2 emission continues up to the year 2100, then the warming in the year 2100 would only be about 60% of the 'committed warming' from the CO_2 concentration in 2100. This calculation seems rather callous, almost sneaky, given the inevitability of warming once the CO_2 is released."
– Professor David Archer, 'How much CO_2 emission is too much?', RealClimate (blog), Nov 6th 2006

but also said a couple of months later that:

> *"In the 20 years that I have been a scientist with the IPCC, I have not encountered a government trying at this stage to influence the assessment beyond making suggestions that would genuinely help its remit or focus."* [196]

There are arguments from all sides as to the importance of this political influence, with some denialists even claiming that it invalidates all of the IPCC's findings and so that anthropogenic climate change is just a political conspiracy.

However, since the scientific draft of the latest SPM was leaked, we can compare it with the published version and see for ourselves. The changes are not massive and outrageous, but any political involvement in the process of generating supposedly authoritative science is dubious, and it appears clear that if anything they lean towards softening the case for urgent action. [197]

Figure 30: The difference between warming by 2100 and ultimate warming, for a given CO_2 concentration. Source: www.ipcc.ch/ pdf/climate-changes-2001/synthesis-spm/synthesis-spm-en.pdf, p.22.

Time horizon of IPCC figures

Quite apart from their political involvement, long drafting process, and poor predictive track record, there is also one other crucial issue with the IPCC approach which is rarely mentioned – much of their work discusses a time-horizon of 2100 (since that is what the UN Framework Convention on Climate Change – of which the Kyoto Protocol is a part – focuses on). So when the IPCC present the necessary emissions targets for,

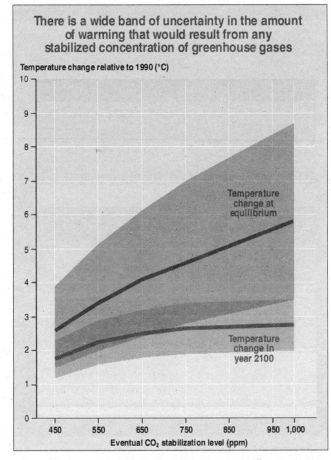

say, restricting global temperature rise to 2 degrees, they are often talking about keeping it to 2 degrees *by 2100*.

Figure 30, taken from the IPCC's 2001 synthesis report, confirms that for a given stabilised CO_2 concentration there is a substantial difference between the resultant temperature change by the year 2100 and the ultimate temperature change, but this is not widely discussed. Unless we plan on maintaining a habitable planet only to the end of this century, we must be careful which of the IPCC's figures we use.

Reasons for considering the IPCC position

Given all this, we might be tempted to question the usefulness of the IPCC's work, but it remains essential for a number of reasons. Firstly, the IPCC's *data* are the most reliable available, as we saw in the 'Climate change explained' chapter (p.130).

Secondly, despite its shortcomings, their modelling work does effectively present a well-supported 'best-case' scenario – given the amount of scientific work incorporated in their final reports, and especially considering the factors we have explored here, it is hard for anyone to convincingly argue that the situation is *less* urgent than their work suggests. As discussed in Part One, one of the key roles of the Transition movement lies in challenging deeply-held cultural assumptions, and robust and widely respected evidence supporting our position

is a critical tool to that end. Regardless of their shortcomings, the IPCC's findings call for a drastic change in direction, and this should help readers to be able to argue for the Transition Vision in their communities without the necessity of first engaging with the arguments about the validity of the IPCC position.

Indeed, it could be argued that the conservative findings of the IPCC are to some extent a necessary result of the role it plays as an authoritative, politically involved scientific body. Whether we agree or not with scientists considering politics when presenting their findings, we can appreciate that if the IPCC was found to be *over*stating the case it could be damaging to the process of ensuring adequate responses.

It is also true that many of the shortcomings of the work are both known to and acknowledged by the IPCC. The scientific work to model the many complexities and uncertainties in the climate system is still very much underway, and so entirely accurate and precise findings are unavailable. Consequently, there is a balancing act here, and the IPCC tend to opt for producing precise figures while acknowledging the factors that are not yet incorporated.

Still, it is clear that the IPCC, with all their caution, are understating the extent of the problem, and for our purposes a precise answer to an incomplete question is of little use. With this in mind, this book will now move on to study our best understanding of the reality of climate change, even where that leads us into areas of greater uncertainty.

Chapter 19

Climate change – a reality check

Climate Code Red

This section of the book draws extensively on the findings of the outstanding *Climate Code Red* report, published by the Carbon Equity group in 2008 to clearly communicate the severity and urgency of our climate emergency. First, let us consider a parallel drawn by the authors between the risks we accept in other areas, and those we are accepting regarding climate change:

> *"For nuclear power stations in the USA, the regulatory standard is that there should be no more that one-in-a million risk of serious accident. In 2004, the chance of being killed in a commercial air crash was about one in four million. If instead the risk was one in a thousand — a 0.1% chance — we would not fly.*
>
> *Yet we seem to accept much higher risks as reasonable in setting global warming targets. The talk is about a 20–30% species loss for a rise of 2°C, very likely coral reef destruction, possible ice-sheet disintegration and the prospects of economic damage 'on a scale similar to those associated with the great wars and the economic depression of the first half of the 20th century' (according to Nicholas Stern) as if it were a game of chance, a poker hand where with an ounce of luck the right cards will be dealt and the Earth will 'get out of jail free'."* [198]

With this in mind it is worth considering that a 'best estimate' emissions trajectory for keeping warming to two degrees effectively represents only a 50% chance of keeping global temperature rise to less than 2°C. In attempting to represent the full range of uncertainties in climate-change projections, the IPCC has downplayed many of the low-probability, high-consequence events. Yet when the stakes extend to the possible end of complex life on Earth, surely the precautionary principle demands that we study these worst-case scenarios very carefully indeed, and bear them in mind at all times?

If we were offered money to point a gun at our head and pull the trigger, we might be less than entirely reassured by the news that the majority of experts consider there is only a 10% chance that is loaded. We would be well advised to avoid the risk altogether. As Dr Hansen put it in 2007:

> "For the last decade or longer, as it appeared that climate change may be underway in the Arctic, the question was repeatedly asked: 'is the change in the Arctic a result of human-made climate forcings?' The scientific response was, if we might paraphrase, 'we are not sure, we are not sure, we are not sure. . . yup, there is climate change due to humans, and it is too late to prevent loss of all sea ice.' If this is the best that we can do as a scientific community, perhaps we should be farming or doing something else." [199]

We need to consider these possibilities *before* we become clear and certain that it is too late. And as we have seen, perhaps the darkest possibility is that of setting off unstoppable runaway global warming. As *Climate Code Red* says:

> "The final and most profound issue when determining how quickly we need to get to a safe-climate condition is how close we are to runaway warming, the circumstance in which there is sufficient momentum from natural positive feedbacks that the warming becomes too powerful for humans to stop, no matter how hard we try. Hansen and Sato have said that the threshold for runaway warming is likely to be a 1.7°C rise above pre-industrial levels." [200]

In other words we are perilously close to losing control of the situation and running out of options altogether. If we allow global temperature to rise by more than 1.7°C, we will likely be committing the planet to the sixth mass extinction in its long history, with humanity very much on the endangered list. As Dr Hansen and colleagues bluntly and chillingly put it in a 2008 paper, it is simply becoming a question of whether "humanity wishes to preserve a planet similar to the one on which civilization developed and to which life on Earth is adapted". And as we saw in the last two chapters, on current trends a 1.7°C rise above pre-industrial levels looks anything but unlikely. [201]

So, to the obvious question – if we were to apply the same precautionary principle to our planetary environment that we do to our own personal safety, what emissions targets would we be striving for?

But there is no obvious answer. The globally agreed aim is to "prevent dangerous anthropogenic interference with the climate system", yet *Climate Code Red* argues that: [202]

> "The primary assumptions on which climate policy is based need to be re-interrogated. Take just one example: the most fundamental and widely supported tenet — that 2°C represents a reasonable maximum target if we are to avoid dangerous climate change — can no longer be defended. Today at less than a 1°C rise the Arctic sea-ice is headed for very rapid disintegration, in all likelihood triggering the irreversible loss of the Greenland ice-sheet, catastrophic sea-level increases and global temperatures rises from the Albedo flip (dark water absorbing more heat than reflective white ice). Many

"We have already passed the stage of dangerous climate change. The task now is to avoid catastrophic climate change."
– Professor John P. Holdren, Chairman of the American Association for the Advancement of Science, August 2006

"It's not something you can adapt to. We can't let it go on another 10 years like this. We've got to do something."
– Dr. James Hansen, 'Debate on climate shifts to issue of irreparable change', *The Washington Post*, January 29th 2006

"I was asked, 'What is the appropriate stabilization target for atmospheric CO2?' " he recalled. "And I said, 'Well, I think it's inappropriate to think in terms of stabilization targets. I think we should think in terms of emissions targets.' And they said, 'O.K., what's the appropriate emissions target?' And I said, 'Zero'."
– Dr Ken Caldeira on briefing members of the US Congress, 'The darkening sea: what carbon emissions are doing to the ocean', *The New Yorker*, Nov 20th 2006

species and ecosystems face extinction from the speed of shifting isotherms. Our carbon sinks are losing capacity, and the seas are acidifying.

If we could start all over again, surely we would say we must stabilise the climate at an equilibrium temperature that would ensure the stable continuity of the Arctic? Given that this safe level has long since been passed, as soon as we knew there was a problem with the climate, we should have aimed for a level of atmospheric CO2 that would allow the restoration and then maintenance of the Arctic ice cap, with a safe margin for uncertainty and error." [203]

In other words, to achieve our agreed goal we would have to produce fewer emissions than we have already. The authors go on to argue, based on recent science, that to actually "prevent dangerous anthropogenic interference with the climate system" we would need to keep the global average temperature to no more than 0.5°C over pre-industrial levels. Yet we have already seen a rise of 0.8°C, and at least another 0.6°C is still due *from emissions to date* – the delay being due to 'thermal inertia', the time it takes to warm or cool the great mass of our planet.[204] But as the authors of *Climate Code Red* state:

"The fact that we have long passed this point in no way detracts from its importance as a policy goal, and a state to which

we should wholeheartedly endeavour to return the planet." [205]

Simply put, the political climate is completely out-of-step with the physical reality. The majority of politicians are still acting as though there were a direct, linear and immediate relationship between emissions and temperature. As a consequence, current negotiations revolve around how to begin *slowing the rate of increase* of emissions in a manner that is 'acceptable to business'.

Meanwhile the reality has moved on – an emergency is upon us. Our current negotiations are akin to standing in a burning shop discussing the slowest acceptable speed at which to leave. Continuing with business as usual in such circumstances – trying on one more pair of trousers – simply represents suicidal insanity.

Let us be clear – achieving a long-term 0.5°C cap would mean returning atmospheric CO2 levels to 300-350ppm, compared with the 2008 level of around 385ppm.[206] If we are truly aiming to avoid 'dangerous climate change' there is no 'acceptable emissions budget' left. As Figure 31 illustrates, to bring temperatures back down we will need to reduce net emissions to less than zero. In other words, the question we should be asking is: 'How quickly can we *reduce* the already-dangerous amount of carbon in our atmosphere?' [207]

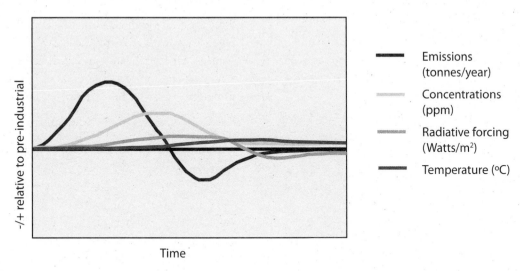

Carbon drawdown

Legend:
- Emissions (tonnes/year)
- Concentrations (ppm)
- Radiative forcing (Watts/m²)
- Temperature (°C)

y-axis: −/+ relative to pre-industrial
x-axis: Time

Figure 31: A scientifically realistic response to climate change.

In such dramatic circumstances it would be easy to argue that we should be asking what carbon concentration reductions are possible, rather than what is acceptable, with the consequences of failure overriding all other concerns, but we have already considered the terrible human costs an instantaneous withdrawal from fossil fuels would entail (p.124).

In the absence of any remaining acceptable carbon budget, our task must be to move away from our carbon-intensive lifestyles as rapidly as possible without allowing the shock of hasty withdrawal to cause intolerable suffering or destroy our ability to co-operate globally. This is the tightrope we face – our true sustainability emergency.

Is reducing concentrations of atmospheric carbon even possible?

The short answer is yes – it has happened many times in the history of our planet, without the need for any human 'high technology'. Dr Hansen highlights that the prevalence of positive feedbacks in our climate system mean that if we could subject the Earth to a modest cooling it would lead to feedbacks that tend to produce further cooling, bringing temperature back towards a more comfortable equilibrium.[208]

There is a huge diversity of proposed methods for achieving such a cooling, with many of them based around the insight that radiative forcing is the driver of temperature, not

"There is always an easy solution to every human problem – neat, plausible, and wrong."
– H.L. Mencken

greenhouse gas concentrations. Accordingly, many suggestions revolve around reflecting the sun's energy in a wide variety of ways, whether by placing mirrors between us and the sun, injecting yet more sulphate aerosols into our atmosphere to block more sunlight, painting roofs and roads white or covering the deserts in reflective plastic.[209]

It is at least conceivable that directly influencing radiative forcing in this way could buy us a little more time as a *supplement* to immediate and dramatic emissions reductions, but the notion that such an approach could allow us to continue to pump out greenhouse gases with impunity is a non-starter. It is not just that we have little understanding of the likely consequences of reduced solar energy input, but that we do have a very good idea of what is happening to our oceans as more and more CO_2 is dissolved into it.

The oceans are already 30% more acidic than they were in pre-industrial times, and according to NASA this increasing acidity has "the potential to cause the extinction of many marine species". Ken Caldeira, a chemical oceanographer at Stanford University warns that, "what we're doing in the next decade will affect our oceans for millions of years. CO_2 levels are going up extremely rapidly, and it is overwhelming our marine systems." This acidification is a most serious problem in itself, quite apart from the fact that a world with dead oceans would not produce a very comfortable climate for us.[210]

Other 'techno-fix' options include creating new 'anthropogenic carbon sinks' (also known as 'synthetic trees'), and there is even a $25m prize on offer – the Virgin Earth Challenge – for anyone who can demonstrate to the judges' satisfaction a commercially viable design which results in the removal of anthropogenic atmospheric greenhouse gases. In the context of peak oil though, such a process would need to use relatively low levels of energy (the carbon-intensity of which would need to be considered), as well as being 'commercially viable'. Clearly, the prize has not yet been claimed.[211]

Perhaps a touch of humility is called for. As Gregory Bateson once wrote:

"The major problems in the world are the result of the difference between how nature works and the way people think."

As we saw on page 142, the natural 'carbon sinks' already draw down huge quantities of carbon from our atmosphere but are declining in effectiveness due to our impacts. Allowing the forests to return to their former glory is just one obvious way in which we could counteract this.

Another would be changing our approach to agriculture. The Woods Hole Research Centre has found that around 25% of carbon build-up in the atmosphere over the past 150 years has come from land use change, mainly deforestation and farming. Ohio University put the figure at around 50%. By avoiding nitrogen fertiliser and building up the soil's organic content this can be greatly reduced (soil naturally contains twice as much carbon as the atmosphere), and with 'no dig' techniques like permaculture even more so.[212]

Australian keyline farming advocate and entrepreneur Allan J. Yeomans argues that this effect is so significant that if the world's soils were managed using design systems like permaculture and keyline, so much carbon would be sequestered that we wouldn't have a climate change problem at all. At a European Commission conference in June 2008, Professor Rattan Lal of Ohio University presented findings that with changes to agriculture and land use, terrestrial ecosystems could naturally reabsorb sufficient CO_2 to reduce current atmospheric concentration by at least 50ppm from current levels (around 385ppm).[213]

As the American journalist Eric Sevareid once said, "the chief cause of problems is solutions". Maybe we need to stop trying to override nature to solve each new set of problems and instead help nature to clean up after us.

There are perhaps two key points to make here. The first is that drawing carbon out of our atmosphere is clearly possible. If we recognise that finding a sustainable solution is the most important challenge on Earth there is little doubt that it can be done. But the second is that to give us that chance, our focus now (and out to 2027 – the full scope of this book) must be on reducing our emissions to virtually zero.

When governments speak of reductions targets for 2050, or of carbon sequestration technologies that will not be ready for twenty years, they might as well be singing lullabies. Technologies clearly have an important role to play, but techno-fixation cannot be the focus of our efforts. We need to act now, with our focus on achieving dramatic emissions reductions over the next few years, as the science demands.

The key to achieving these reductions lies at the community scale. This may seem counter-intuitive, as we are so often told that:

> "Large-scale problems require large-scale solutions."

But while we are often encouraged to think about climate change as requiring national and international level solutions beyond our influence, it is easy to forget that it is people who are generating these emissions, and we *all* live at the local level. In this sense climate change is not so much a global problem as a collective local problem. As David Fleming has pointed out:

> "Large-scale problems do not require large-scale solutions; they require small-scale solutions within a large-scale framework." [214]

We clearly do need binding national and international agreements that we will all do our part, but the actual *doing* is primarily a local matter. Parts One, Two and Three examined the detail of how we can do this and as we saw, the changes we need to make in response to our climate crisis could also benefit us in many other respects.

Isn't the scale of the problem terribly depressing?

This is certainly all challenging information, but really it is just one more symptom of the unsustainability of our current global way of life. It represents all the evidence and

"I am always surprised when people get depressed rather than energised to do something. It's not too late to stabilise climate. . . I am not about to give up."
– NASA's Dr James Hansen, 'Scientists Hopeful Despite Climate Signs', Associated Press, September 23rd 2007

motivation we need to declare transition to be our utmost priority – the first aid for our global state of emergency.

Yet it is more than that. In facing the reality of the situation we can begin to see it for what it is, rather than what we might fear it to be. At times there can be a temptation to despair, and in total contrast to the teachings of our culture the deep ecologist and Buddhist scholar Joanna Macy suggests that we allow ourselves to feel it fully:

> *"The suppression of despair, like that of any deep recurring response, contributes to the numbing of the psyche. . . . Of all the dangers we face, from climate chaos to permanent war, none is so great as this deadening of our response. For psychic numbing impedes our capacity to process and respond to information. The energy expended in pushing down despair is diverted from more crucial uses, depleting the resilience and imagination needed for fresh visions and strategies . . .*
>
> *It is good to realize that falling apart is not such a bad thing. Indeed, it is as essential to transformation as the cracking of outgrown shells. Anxieties and doubts can be healthy and creative, not only for the person, but for the society, because they permit new and original approaches to reality. . .*
>
> *Many of us fear that confrontation with despair will bring loneliness and isolation. On the contrary, in letting go of old defences, we find truer community. And in community, we learn to trust our inner responses to our world—and find our power."* [215]

The Transition movement itself is growing out of this philosophy of facing challenges together and unearthing our communal power. Facing reality (and our true feelings about it) becomes the first step towards creating the happy, thriving future we want to see, in partnership with all the other individuals, groups and organisations working in the same direction.

What is truly depressing – even terrifying – is continually being told that the supposed 'worst case' worsens with every assessment. How can we address a problem realistically until we receive an honest assessment which doesn't hold out false hopes?

We need to know the scale of the problem, and the real best- and worst-case scenarios, before we can reasonably decide on how we are to respond to it. Someone told me recently that "the one thing you can be sure of when someone talks to you about climate change is that they believe it's worse than they are letting on." There is some truth in that, but it doesn't apply here – this is our unvarnished best understanding.

And that puts us back in the realm where happy surprises may be lurking. At present our actions are simply overwhelming Nature's ability to compensate for our emissions, but if we can begin to control our excesses we may well find that she is more resilient than our best assessment has given her credit for. [216]

And crucially, while writing this book I have become convinced that even taking into account all of the most challenging evidence, we can still work together, locally and globally, to leave a legacy of which to be proud.

Part Five

UK CONTEXT

The UK Government's May 2007 Energy White Paper stated that:

"Energy is essential in almost every aspect of our lives and for the success of our economy. We face two long-term energy challenges:

• tackling climate change by reducing carbon dioxide emissions both within the UK and abroad; and

• ensuring secure, clean and affordable energy as we become increasingly dependent on imported fuel." [217]

In Part Five we will consider whether the Government is acting on this understanding, and whether they are recognising that while these are indeed long-term challenges, they are also very immediate ones.

Chapter 20

Peak oil in the UK

UK Government position

Having examined the global context of peak oil in Part Four, you may well be wondering how our Government is responding to this challenge. A petition presented to Downing Street in 2007 asked the Prime Minister to acknowledge the situation of peaking oil and natural gas supplies and to take "immediate and responsible action" to address it.

In October 2007 the Government issued a response, which read in part:

> *"The Government fully recognises that there is uncertainty around the issue of future global oil and gas production. However, on the balance of the available analysis and evidence, the Government's assessment is that the world's oil and gas resources are sufficient to sustain economic growth for the foreseeable future.*
>
> *As the International Energy Agency (IEA) notes in its 2005 publication 'Resources to Reserves: Oil and Gas Technologies for the Energy Markets of the Future', the key problem is not the limit of geological oil resources."* [218]

There are two reasons why our Government's reliance on this IEA report fails to reassure. Firstly, the IEA have not proved a reliable watchdog. If we look to the IEA's 2004 World Energy Outlook, which was the latest when *Resources to Reserves* was published, we find the following projection:

> *"The average IEA crude oil import price, a proxy for international prices, is assumed in the Reference Scenario to fall back from current highs to $22 in 2006. It is assumed to remain flat until 2010, and then to begin to climb in a more-or-less linear way to $29 in 2030."* [219]

The report does also state that:

> *"In view of the uncertainty surrounding oil prices, we have carried out a separate analysis to examine the effects of high oil prices on world oil supply and demand, OPEC revenues and global oil investment."* [220]

But even this "high oil price" scenario only considered oil prices averaging $35 per barrel (in year-2000 dollars) until 2030. The IEA completely failed to foresee the scale of price increase we saw, with oil prices reaching more than $140 in July 2008.

The second cause for concern is that the comments coming out of the IEA itself suggest that they have now finally seen the writing on the wall, and that their position has shifted substantially since the 2005 report in which our Government still places such faith. The IEA's Chief Economist, Fatih Birol, has said that he has since experienced "an earthquake" in his thinking. In June 2007 he said that: [221]

> "If Iraqi production does not rise exponentially by 2015, we have a very big problem, even if Saudi Arabia fulfils all its promises. The numbers are very simple, there's no need to be an expert." [222]

With the continuing instability in Iraq, nobody is expecting exponential increases in oil production from there, so Birol's words were a thinly veiled warning. In March 2008 Birol spoke with more transparency and urgency:

> "We are on the brink of a new energy order. Over the next few decades, our reserves of oil will start to run out and it is imperative that governments in both producing and consuming nations prepare now for that time. We should not cling to crude down to the last drop – **we should leave oil before it leaves us**. . . The really important thing is that even though we are not yet running out of oil, **we are running out of time**." [my emphasis] [223]

In April 2008 he re-emphasised the point, saying:

> "We sounded the alarm bells in Nov. 2007 and this November, in the World Energy Outlook 2008 report the bells may well shrill much louder... It is up to the governments, we have warned them." [224]

The IEA's World Energy Outlook 2008 report duly opened with the following blunt statement:

> "The world's energy system is at a crossroads. Current global trends in energy supply and consumption are patently unsustainable – environmentally, economically, socially." [225]

Our Government's position has been becoming ever more untenable. In May 2008 our Government's latest projections were published, with a central scenario giving oil prices reaching $70/barrel by 2020, despite the fact that the price was already at $120/barrel. Yet later that same month Prime Minister Gordon Brown finally acknowledged the reality, admitting that: [226]

> "The cause of rising prices is clear: growing demand and too little supply to meet it both now and – perhaps of even greater significance – in the future. . . . Our strategic interests – reducing energy costs, increasing our energy security, tackling climate change – all now point in the same direction: decreasing dependency on oil." [227]

In July 2008 he went further, saying that:

> "We must now leave behind the old wasteful, oil-dependent ways of yesterday and embrace the new cleaner and sustainable energy future of tomorrow. . . . [We must] set ourselves on a new energy path – a path from our economies that are today over-dependent on oil towards the post-oil energy economies of the future. . . . Today our globalised, energy-hungry and warming world requires a shift from oil dependence

"A spectre is haunting Europe – the spectre of an acute, civilisation-changing energy crisis."
– Jeremy Leggett, former oil geologist and member of the UK Government's Renewables Advisory Board, 'What they don't want you to know about the coming oil crisis.' *The Independent on Sunday*, January 20th 2006

to sustainable energy. . . . Only with political leadership from all of us will we be able to move towards a new sustainable economy. This is now Britain's goal." [228]

These comments are a hopeful sign, and we must ensure that they are followed up by action, for if the Government continues to assume business as usual, we could be walking blindly into disaster. Of particular concern is that these comments seem to have been prompted by high oil prices, rather than by an understanding of the underlying realities, and so as prices fluctuate Government resolve may prove equally unreliable (see p.123).

As we will see below, natural gas supply is also likely to present serious challenges here in the UK in the near future, but the Government's attempts to increase resilience have thus far been deeply inadequate.

UK oil and natural gas production trends

Our Government's apparent complacency over recent years has been even harder to understand when we consider that the UK's own oil and gas production peaked in 1999 and has been in steep decline since. Figures 32 to 34 are from the Government Department for Business, Enterprise and Regulatory Reform (BERR, formerly the DTI).[229]

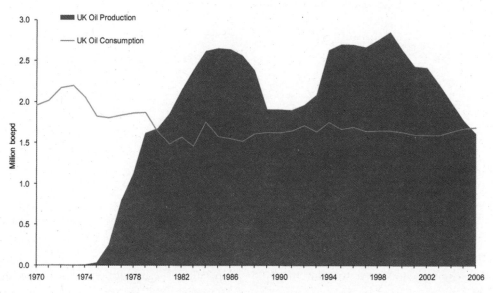

UK Oil Production and Consumption

Figure 32: UK oil production and consumption.

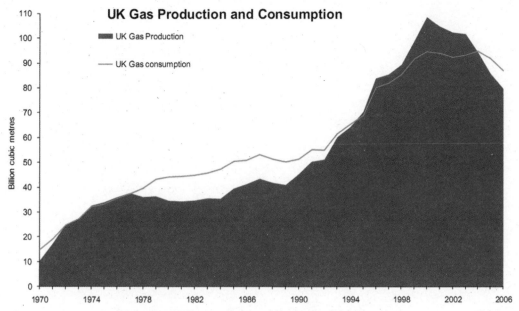

Figure 33: UK natural gas production and consumption.

Future implications

In 2007, around three quarters of the primary energy consumed in the UK was derived from oil and gas. Until recently we were self-sufficient in these fossil fuels, and were even profiting from exporting oil and gas as recently as 2003 (at the relatively low prices of that time), but now we are reliant on ever-increasing imports.

With current trends and policies, the Government predict that by 2010 we could be importing a third or more of the UK's annual natural gas demand. By 2020, we could be importing around 80% of our gas (and 75% of our coal). Because of these trends, the cost to the UK of importing its oil and gas needs alone could be set to grow to about £100 billion *per year* by 2013. To put this into context, since the first year of 'surplus'

(oil exports exceeding oil imports) in 1980, oil trade has contributed £86 billion *in total* to the UK balance of payments. The largest annual 'surplus' was £8 billion in 1985. [230]

In reality it is unlikely that the UK will be able to find both the money and the willing sellers to enable us to import such amounts – our wealthy status means that we have so far been relatively insulated from oil demand destruction (p.127), but this cannot last forever with clearly limited amounts of exported oil available on the world markets.

Global 'peak oil exports' is likely to come before 'peak oil production' as oil-exporting countries reap huge profits from increasing oil prices, resulting in growth in their economies and populations, and thus in increased domestic oil consumption. This means that the amount of oil left over for export declines more rapidly than the total production.

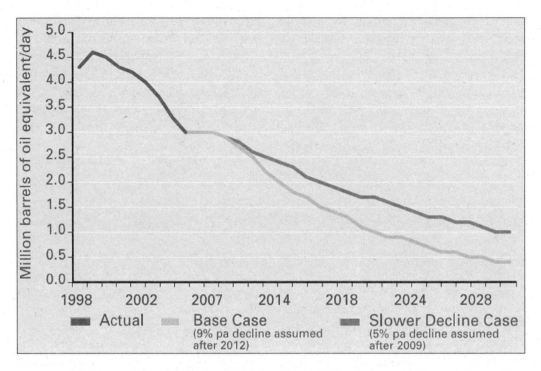

Figure 34: Government projection for UK continental shelf oil and gas production (combined).[231]

Jeffrey Brown and Samuel Foucher have examined this issue closely and conclude that not only are global oil exports *already* in decline, but the current top five oil exporting countries – Saudi Arabia, Iran, Russia, Norway and the United Arab Emirates – are likely to provide *zero* net oil exports by about 2031. Barring friendly extra-terrestrials, there is no-one else left to make up the deficit.[232]

However, while we have already examined many of the challenges oil supply disruption is likely to bring (p.124), here in the UK peak gas may be our more immediate concern.

Few countries use natural gas for such a high proportion of their primary energy consumption as the UK (38%, compared with 25% in the US for example). We use it for heating and cooking as well as for generating electricity, so serious shortages would be extremely disruptive, and, crucially, it is much more difficult and expensive to transport than oil, with major infrastructure required for imports. In 2007 our gas imports rose by 39% compared with the previous year – meaning we are already relying on imports for more than a fifth of our natural gas demand – yet the sources on which we are becoming ever more dependent do not look reliable.[233]

Our Government's strategy for securing additional natural gas imports is twofold. The first aspect is our pipeline infrastructure to Europe, including the new Langeled pipeline to Norway, which is large enough to supply the majority of the projected UK import requirement in 2012, *if* someone is willing to send gas through it. But here the extent of the pipeline network means we are effectively in a closed market, which not only limits the total available supply, but also means we are competing with European and Russian demand for our share – we are bidding against prosperous competition in our attempts to buy our way out of the problem.

In addition, in May 2008 the Executive Vice-President of the Norwegian pipeline company Gassco said bluntly that regardless of the price we might be prepared to pay, long-term contracts with mainland Europe mean that for Norwegian gas exports: [234]

> *"The UK is a secondary priority. . . like it or not, that is a fact."* [235]

The other half of our natural gas import strategy is the Liquefied Natural Gas (LNG)[236] terminals we have built in order to source gas from further afield, on the assumption that we can find countries with spare capacity that they are willing to sell at an affordable price. The problem is that globally there is twice as much LNG *import* capacity as there is LNG *export* capacity – it is a seller's market. We had a clear demonstration of this when in the winter of 07/08 Japan and Turkey were offering twice as much as Britain for LNG imports, and so, in accordance with the law of the market, our LNG terminals sat unused while deliveries were diverted there, sometimes when actually *en route* to Britain.[237]

So as the economic reality of our trade deficit and the physical reality of limited supplies bite it is very likely that we will see natural gas supply shortages within the UK. This means that without an energy rationing system such as TEQs, the poor will be forced to go without reliable energy supplies, while a far greater proportion of most people's disposable income will be swallowed up by energy purchases.[238]

Any unusually cold winters would mean more gas demand in homes and power plants than our liberalised energy market – or its infrastructure – might be able to supply. This would have direct human impacts, as already there are around 25,000 'excess winter deaths' each year in the UK, with most attributable to old people not being able to keep warm enough.[239]

Our options for increasing supply are increasingly limited and desperate, so we must learn to reduce demand *before* crisis hits. It is yet another reason why we must reduce our dependence on fossil fuels.

"Look at South Hook, where you have these new (LNG) facilities that are going to be capable of taking 15.6m tonnes a year. I reckon, realistically, we're not going to see more than about half of that come to the UK. The rest will be sent to meet higher spot prices all over the world."
– Frank Harris, head of LNG at the Wood Mackenzie energy consultancy, 'Price war threat to UK gas supplies', *The Sunday Times*, March 9th 2008

"I don't want to exaggerate, but gas comes from some fairly unstable parts of the world and some not readily associated with human rights and democracy."
– UK Energy Minister Malcolm Wicks, 'Wicks: All is lost on global warming without clean coal', *The Guardian*, August 8th 2008

Chapter 21

Climate change in the UK

UK Government position

As the quote from Tony Blair (see left) indicates, the UK Government have been considerably more forthcoming in speaking about the threat of climate change than about peak oil. They have been vocal in pushing the issue forward internationally, and promised in three consecutive election pledges to cut UK carbon dioxide emissions by 20% (from 1990 levels) by 2010.

The Government made great play of the fact that this figure is considerably stricter than Britain's Kyoto Protocol obligation of 12.5% cuts in greenhouse gases (from 1990 levels) by 2012. The Kyoto targets will likely be met (due largely to the country's power stations switching from coal to natural gas in the 1990s for other reasons) but it is now almost certain that the Government's own target for CO_2 will not be met. Since 1997 CO_2 emissions have been rising slightly. According to Government figures, CO_2 emissions rose by 1.2% in 2006 compared with 2005, an increase of 6.4m tonnes, and the reduction in CO_2 since 1990 has been just 5%.[240]

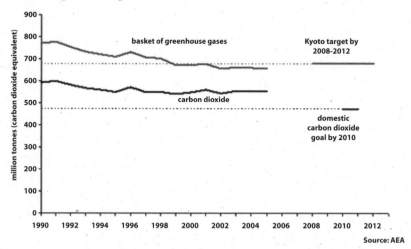

Emissions of greenhouse gases 1990-2005

Figure 35: UK greenhouse gas emissions versus targets.[241]
Source: http://www.defra.gov.uk/news/2007/070131a.htm

At the launch of the Government's Climate Change Programme review in March 2006 the then Environment Secretary Margaret Beckett admitted this, saying:

"I regret the fact that we have not identified in this programme the measures to get us down to 20% [reduction in CO_2 emissions]. It has proved to be a more difficult task than we had hoped to reach the targets that we had originally set." [242]

The reason it is proving so difficult was summed up recently by the most unlikely of speakers:

"The fact is that a low carbon society will not emerge from thinking of business as usual." – UK Prime Minister Gordon Brown, June 2008 [243]

We can only hope that this time our Government mean what they are saying.

As has been widely publicised, Parliament has produced a Climate Change Act which became law in November 2008, and which is the first legally binding long-range national carbon-reduction target. The Act originally mandated 60% emissions cuts (from 1990 levels) by 2050, and by 26-32% by 2020, and has since been revised to 80% cuts by 2050.[244]

However, there are two fundamental problems here. Firstly, as the UK Parliament's Environmental Audit Committee commented in October 2007:

"The organic process by which leadership and responsibility have evolved appears to have created a confusing framework that cannot be said to promote effective action on climate change." [245]

In other words, there is no coherent structure in place to ensure that these new targets are achieved. The Government's track record is not good, and their original aim of cutting emissions levels by around 30% by 2020 with current policies was politely described by another study as "very optimistic".[246]

Secondly, even if we did achieve such targets, they are out of keeping with the scientific understanding of the scale of the challenge. The UK Government regularly leans on the work of Lord Nicholas Stern in this area. Stern is a former Chief Economist of the World Bank and his 2006 *Review on the Economics of Climate Change* (commissioned by Gordon Brown when he was Chancellor) sensibly advised that acting now on climate change would be a lot cheaper than acting later. Unfortunately, the report's summary of the science of climate change was a touch misleading. A widely quoted passage in the Summary of Conclusions says that,

"The risks of the worst impacts of climate change can be substantially reduced if greenhouse gas levels in the atmosphere can be stabilised between 450 and 550ppm CO_2 equivalent (CO_2e). The current level is 430ppm CO_2e today, and it is rising at more than 2ppm each year." [247]

But as we have seen (pp.130-134) this statement conflates two different meanings of CO_2e, with CO_2e(Total) being used for the proposed

"We underestimated the risks . . . we underestimated the damage associated with temperature increases . . . and we underestimated the probabilities of temperature increases."
– Lord Nicholas Stern on the Stern Review, 'Stern takes bleaker view on warming', *Financial Times,* April 16th 2008

"With an atmospheric CO_2 stabilisation concentration of 550ppm, temperatures are expected to rise by between 2°C and 5°C."
– DEFRA, The Scientific Case for Setting a Long-Term Emission Reduction Target (2003), quoted in 'Giving Up On Two Degrees', George Monbiot, May 1st 2007

"A political promise to do something 40 years from now is universally ignored because everyone knows it's totally meaningless."
– Al Gore, 'Al Gore Lays Down Green Challenge to America' , *The San Francisco Chronicle,* July 18th 2008

stabilisation levels, but CO_2e(Kyoto) being used for the current level. These two different concepts cannot be directly compared in this way, and the quoted passage thus gives the misleading impression that we are only 20ppm from breaching the 450ppm stabilisation target. In fact CO_2e(Total) was still at around 375ppm in 2005 (as explained from p.132). As the Stern Review is such a high-profile document this has unfortunately spread a deal of misunderstanding and confusion.[248]

As we explored in Chapters 18 and 19, the 450-550ppm stabilisation target discussed in the Review is also substantially higher than scientific realism demands. The Government's original '60% reductions by 2050' target was set on the basis that it would, if part of a commensurate global effort, keep CO_2 concentrations to 550ppm, with the aim of keeping global mean temperature rise to less than two degrees. Yet the IPCC table on page 139 shows that 550ppm would lead to almost four degrees of global temperature rise – a catastrophic increase – and as discussed from page 140 onwards, there are good reasons to believe they may even be *understating* the case.[249]

This is all important information, yet it is actually not ultimate stabilisation levels that are the critical topic, nor what we say we will do by 2050, but rather what we start doing today, and tomorrow, and for the next few years.

It is worth noting that there are some countries trying to urgently address the problem at the political level – President Oscar Arias of Costa Rica, for example, argues that a climate-neutral economy is a competitive economy, and has unilaterally declared that his country will become completely carbon-neutral by 2021, as part of his 'Peace with Nature' presidential initiative (perhaps a somewhat more inspiring concept than a 'War on Drugs').[250]

Yet in Britain many politicians still argue that there is not enough public will for realistic targets on climate change, or for a framework like TEQs that would guarantee that the targets we set are actually achieved. This is one reason why the Transition movement is so important – its rapid spread shows politicians that we are concerned and acting on that concern, which in turn helps embolden them to take stronger political action.[251]

Still, as discussed in previous chapters, regardless of the politics, lags in the climate system mean that we are already committed to the climate change we are likely to see in the UK by 2027, so let's examine the likely impacts, starting with a look at those that are already taking place.

UK climate trends

As we have seen, climate change is a complex matter, and it does not impact all areas of the world equally. It will bring droughts to some areas and floods to others, and the poles will warm more quickly than the equator, yet the trend is towards greater extremes and a general shift away from the 'normal' climate to which we have adapted.

The key body examining the direct impacts of UK climate change is the UK

Climate Impacts Programme, based at Oxford University, who provide many useful and usable resources on their website. Their latest report (UKCIP08) has been delayed until Spring 2009 and will be important reading for Transitioners interested in seeing their local outlook, but a number of significant national trends are already apparent.[252]

The longest available instrumental record of temperature in the world – the Central England Temperature (CET) series – goes back to 1659 and shows that average temperatures in the UK have increased by around 1.5°C since pre-industrial times, with around 1°C of that rise measured just since the 1970s.

Nine of the ten warmest years since 1659 have occurred since 1989, making the 1990s the warmest decade in central England since records began, with an average temperature of 10.1°C. The 2000s (to the end of 2007) are averaging over 10.4 degrees, and the 19.7°C average in July 2006 was the hottest month yet recorded.[253]

This change in our climate has already had a number of effects. Heatwaves have become more frequent in summer, while there are

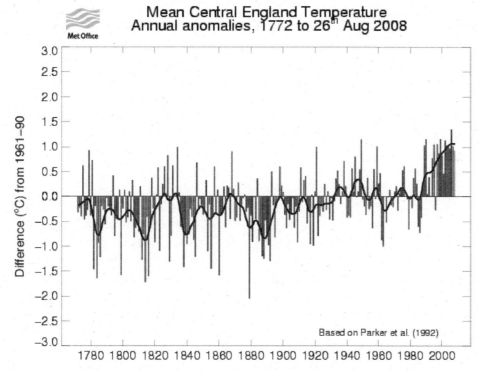

Figure 36: UK historical average temperatures.
Source: http://hadobs.metoffice.com/hadcet/ [254]

"In 2003 we had the European heatwave which was estimated to cost 35,000 people their lives. . . on long-term trends this is something you would expect to happen every 20,000 years, but (if we continue business as usual) this will be an average summer event in 2040, and will be unusually cool by 2060. This is not science fiction any more."
– Professor John Schellnhuber, founding Director of the Potsdam Institute for Climate Impacts Research and the German Government's Chief Advisor on climate change, 'Climate Change – The Short View', BigPicture.tv, Dec 11th 2006

now fewer frosts and winter cold spells. The growing season for plants in central England has lengthened by about one month since 1900, yet an unfamiliar and unreliable climate is a friend to neither plants nor growers. Mild winters, warm early springs and the direct effects of rising CO_2 concentrations in the air have all been shown to increase plants' vulnerability to damaging late-season frosts, and frost-damaged plants, in turn, become less resistant to periods of drought.[255]

Annual mean precipitation (rain and snow) over England and Wales has not changed significantly since records began in 1766, but over the last 200 years winters have become much wetter relative to summers throughout the UK. Over the past 45 years all regions of the UK have also had less even winter rainfall which is more focused in days of heavy precipitation. Severe windstorms around the UK have also become more frequent in the past few decades, though not above what was seen in the 1920s.

Meanwhile, the warming over land has been accompanied by warming of UK coastal waters, with sea-surface temperatures around the UK coast having risen by about 0.7°C over the past three decades, and the average sea level around the UK (after adjusting for natural land movements) about 10cm higher than it was in 1900, with the rate of rise in the 1990s and 2000s increasing due to the faster warming and melting ice around the world. These changes are causing bigger waves (thus increasing coastal erosion) and more severe storms, as well as profoundly disrupting natural ecosystems.[256]

It is perhaps worth mentioning here the oft-heard refrain that "proving a direct causal link between a specific weather event and climate change is impossible". This is true, just as one cannot prove beyond doubt that a given driving accident was caused by blood-alcohol levels. Yet while there is no way to prove that an individual driver wouldn't have crashed if they'd been sober, we have overwhelming evidence that drinking and driving causes accidents. The same is true of extreme weather events caused by climate change.[257]

Future implications

The 2008 UK Climate Impacts Programme report (UKCIP08) will provide the latest set of scenarios for 21st-century UK climate change when it is released, but their previous report in 2002 confirmed that the existing trends are only likely to continue and accelerate.[258]

The short-term trends and projections are bad enough, but the real key is the post-2027 effects both in the UK and around the world that are 'locked in' by our choices before 2027. As we have seen in Part Four, the years we are now living are the time when the future of our planet's climate for millennia to come will be decided.

Closing thoughts

There will be those who argue that the analysis in this book is too dark – that it overstates the challenges that we face – and there will be others who argue that it is too optimistic and too late. Of that much we can be sure.

But I agree with Richard Heinberg, who wrote recently that:

"All that matters in the final analysis is whether awareness leads to effective action that actually reduces the risk of worst-case scenarios materializing."

We could waste all our time and energy analysing probabilities and debating, but far better to spend that time doing. As we engage with our challenges on a practical level we not only improve the situation (and even in the most desperate situations there are better and worse courses of action), but, perhaps even more crucially, we help ourselves.

Humans operate on the human scale, and planting a tree, tending a garden, helping a neighbour – these things do our soul good. And when our minds dwell on the fear that the situation is too overwhelming for us alone to face, they can be soothed by the knowledge that we are contributing to the Transition movement which is swelling all around us. In community we find strength, and all our thoughts and beliefs are somehow hollow until they find expression in action.

If I was born into a world so damaged as to be doomed, then I choose to be a man who fights against that fate with all my will and creativity. And if, as this analysis has further convinced me, I was born into a world of possibilities, then I choose to be a man who plays the biggest part he can in assuring humanity's evolutionary leap to sanity, sustainability and joy. At the deepest level, it doesn't really matter to me whether or not it is *probable* that we succeed. As Tom Atlee has written:

"Probabilities are abstractions. Possibilities are the stuff of life, visions to act upon, doors to walk through."

If nothing else, I believe this Timeline has shown that the Transition Vision is a possibility – the stuff of life. Ultimately, we all have to decide what our own life is for. I will die one day, and, whether my day is near or far, I choose to look back on my time and know that I walked through the doors which inspired me.

Shaun Chamberlin
www.darkoptimism.org

"When we stop distracting ourselves by trying to figure the chances of success or failure, our minds and hearts are liberated into the present moment. This moment then becomes alive, charged with possibilities, as we realise how lucky we are to be alive now, to take part in this planetary adventure." [259]

"The best time to plant a tree was 20 years ago. The next best time is now."
– Chinese proverb

"If we don't fight hard enough for the things we stand for, at some point we have to recognize that we don't really stand for them."
– Paul Wellstone

2009

2030

Ongoing process / feedback

This book is intended as the start of an ongoing process, responding to the changing needs of the Transition initiatives and the developing situation in the UK and wider world. Comments and feedback are welcomed through the Transition Forums at:

http://transitiontowns.org/forum/forum.php?id=22

We are also considering writing the next edition of this book collaboratively through Appropedia, as is currently being done with *The Transition Handbook* at:

http://www.appropedia.org/The_Transition_Handbook

Suggestions have already been received for additional sections of Part Two on the following topics: education, the arts, biodiversity, buildings/housing, resource management (plastics, human waste etc.), communications, manufacturing and industry, land management (including forestry), defence/military and politics (local/regional/national). A fuller, dedicated section on economics would also be an important addition. Any comments on this list of topics, or eagerness to contribute an exploration of them will be welcomed on the Transition Forums detailed above.

This book has not attempted to quantify the energy/emissions footprint of each aspect of the Transition Vision, but this represents a critical avenue for further work.

"Don't ask what the world needs. Ask what makes you come alive, and go do it. Because what the world needs is people who have come alive."
– Howard Thurman

Appendix A

Substitution problem calculation

As outlined on p.128, when we consider the interplay of peak oil and climate change we encounter what is widely known as the 'substitution problem' – as the availability of relatively clean (low emissions) and useful (high EROEI)[260] liquid fuels declines, we are liable to turn to dirtier, less useful energy sources.

It is a common concern that the exploitation of such dirty alternative fuels (such as the Alberta tar sands) may cause our total greenhouse gas emissions to increase, even as emissions from conventional oil and natural gas production decline in line with depletion. However, in this appendix we will make the (perhaps counterintuitive) case that the scale of probable oil production declines between 2009 and 2027 means that *even the maximum feasible expansion of alternative liquid fuels would not maintain current emission growth rates.*

A back-of-the envelope calculation reveals that it is simply not feasible to bring enough projects on stream to make up the shortfall, even if maximising production of these inferior liquid fuels were deemed a worthwhile project.

We are liable to see a peak in global oil production in the next few years. If we consider a peak in 2010 at 90m b/d, followed by a relatively gentle 3% decline rate, we would have a production shortfall of almost 20m b/day by 2018, relative to the peak.

According to Alex Farrell, a professor in the Energy and Resources Group at the University of California who studies the impacts of unconventional oil, tar sands emissions (from production and use combined) are 15-40% higher than conventional oil, coal-to-liquids emissions are almost double those of conventional oil, and for oil shale they can even be more than twice as high. Consequently, let's generously allow for double the emissions per barrel from all unconventional sources. We would still need production of 10m b/day by 2018, and 18m b/day by 2027 to make up for the drop in emissions levels caused by peak oil.[261]

But according to experts like Ray Leonard and David Hughes, even if we were to ignore the environmental impacts and increase unconventional oil production as much as possible, the complex nature of such production imposes challenges of its own. Limits on the availability of water, natural gas, pipelines and suitable personnel mean that unconventional production of bitumen, tar sands and oil shale combined is likely only able to increase by around 2m b/day by 2018, with a 3.7m b/day increase looking the limit of possibility.[262]

Similarly, Robert Hirsch's team estimate that a *worldwide crash programme* of building coal-to-liquids (CTL) plants could hope to bring the first plants on stream after four years, and add ½m b/day in capacity each year from then on. So if this programme was started in 2009, the highest additional contribution CTL could provide would be 3m b/day by 2018.[263]

Finally, biofuels production (currently 1.3 million barrels/day globally) is unlikely to be able to make substantial production increases by 2018, as it is already facing fundamental challenges of land availability and conflict with food production.[264]

So even taking generous figures throughout, the maximum achievable substitution of alternatives appears likely to produce only 5m b/day of additional liquid fuels over the next ten years, generating at most half the emissions levels lost to peak oil.[265]

Nonetheless, two things must be emphasised at this point – that the drop in global emissions caused by oil and natural gas depletion will *not* be sufficient to address climate change (as explored in detail in Part Four), and that nothing here suggests that a total global emissions increase is impossible by 2027, only that the direct production of substitute liquid fuels cannot counteract the emissions-reducing effects of oil depletion.

Global emissions could well increase if, for example, we trigger more serious climate feedbacks, allow the drying Amazon forest to continue releasing the estimated 90 billion tonnes of carbon it contains, attempt to burn all the remaining coal, continue allowing peatlands to be burnt (perhaps to make way for a small expansion in biofuels production), or further disturb the trillions of tonnes of methane hydrates on the ocean floor (perhaps in a desperate attempt to harness them as fuels).[266]

Appendix B
The Transition Timeline's relationship with Zero Carbon Britain

The Transition Vision discussed in Part Two is similar in many respects to the scenario explored in the *Zero Carbon Britain* (ZCB) report, with both examining the UK outlook to 2027. [267]

The authors of that report were aware of the *Transition Timeline* (TT) project from its inception and have been strong allies, and there was a desire from all concerned that the two projects should complement each other as far as possible and try to avoid duplication of effort.

It is clear that there are a number of important similarities between the two reports in both their assumptions and their aims. For those who have examined both, these should be clear enough, so here I will focus on the significant differences Tim Helweg-Larsen (lead author of ZCB) and I identified:

1. ZCB is a resource primarily aimed at politicians and policy-makers, whereas the TT is designed primarily for use by EDAP teams within Transition initiatives.

2. ZCB focused on recommended national/government level action. The TT provides more detail on the current global context, and looks at the changes 'on the ground' that are likely to directly affect communities.

3. ZCB examines the use of technology and TEQs to increase efficiency and reduce energy demand in order to maintain current standards of living (as conventionally defined). The TT incorporates these approaches, but also explores deeper changes to our cultural stories such as voluntary simplicity, relocalisation and different measures of wellbeing.

4. ZCB quantifies the emissions from each sector of the UK economy, whereas the TT currently provides an overview vision.

5. ZCB did not consider in detail the cultural change necessary to see these visions actually implemented, whereas this is a focus of the TT.

Most importantly, while there will always be differences, we see the two visions as essentially aligned, especially given the inherent uncertainties of looking to the future.

Further reading, references and notes

All references below are available online at: www.darkoptimism.org/timelinerefs.html

Further reading

The references on the following pages provide a wealth of further reading on the areas directly covered, but a carefully compiled list of key additional resources can be found at:
http://www.darkoptimism.org/links.html

References and notes

Acknowledgements and context

1. *Zero Carbon Britain*, Centre for Alternative Technology in collaboration with the Public Interest Research Centre, Tim Helweg-Larsen and Jamie Bull, 2007, www.zerocarbonbritain. org. The relationship between Zero Carbon Britain and the Transition Vision outlined in this book is discussed in Appendix B.

Climate Code Red: The Case for a Sustainability Emergency, David Spratt and Philip Sutton, Carbon Equity/Friends of the Earth Australia, 2008, http://tinyurl. com/65hk7b.

Future Scenarios, David Holmgren, 2008, www.futurescenarios.org.

Introduction

2. This vision of our possible future was not vastly dissimilar from the 'Denial' vision detailed in Chapter 2.

3. Rob Hopkins, *The Transition Handbook: From Oil Dependency to Local Resilience*, 1st ed. (Green Books, 2008), available from http://www.transitionculture.org.

Climate change – a summary

4. It should be noted that despite widespread misunderstanding of its meaning, the total CO_2e (equivalent CO_2) figure is currently estimated to be slightly *lower* than that for simple CO_2. This is because total CO_2e accounts for all the factors that affect global temperature change, some of which are cooling factors. This is explained in detail from p.130.

5. From IPCC Press Release, 18th Sept 2007, http://tinyurl.com/2chmus.

Full report: *Climate Change 2007: Impacts, Adaptation and Vulnerability. Contribution of Working Group II to the Fourth Assessment Report of the Intergovernmental Panel on Climate Change*, M.L. Parry, O.F. Canziani, J.P. Palutikof, P.J. van der Linden and C.E. Hanson, Eds., Cambridge University Press, Cambridge, UK, 7-22. http://www.ipcc-wg2. org/index.html.

Peak oil – a summary

6. For a clearly written exploration of the science of peak oil, building up from the basics, see 'The Science of Oil and Peak Oil', Gail Tverberg, http://tinyurl.com/colwr8.

Chapter 1: Why cultural stories matter

7. For further thinking on the nature and importance of cultural stories see: 'Knowing Only One Story', John Michael Greer, *The Archdruid Report*, May 24th 2006, http:// tinyurl.com/2us2vz.

8. 'Seize the Day – Threshold Moments and the Hope for Change', Sharon Astyk, *Casaubon's Book*, Feb 14th 2008, http:// casaubonsbook.blogspot.com/2008/02/seize-day-threshold-moments-and-hope.html.

9. *The Shock Doctrine: The Rise of Disaster Capitalism*, Naomi Klein, Penguin Books Ltd (2008).

Chapter 2: Vision 1 – Denial

10. CIA World Factbook, national current account balances: https://www.cia.gov/ library/publications/the-world-factbook/ rankorder/2187rank.html.

11. See e.g. 'Response of Nebraska Sand Hills natural vegetation to drought, fire, grazing and plant functional type shifts as simulated by the century model', Mangan, J., *et al.*, 2004, *Climatic Change*, 63, 49-90.

Chapter 4: Vision 3 – The Impossible Dream

12. For more information on Tradable Energy Quotas (TEQs) and its progress towards implementation in the UK see: http://www.teqs.net/. The scheme is also discussed in more depth on p.66.

Chapter 5: Vision 4 – The Transition Vision

13. A first look at the Transition Vision comprises Chapter 8 of Rob Hopkins' *The Transition Handbook: From Oil Dependency to Local Resilience*, 1st ed. (Green Books, 2008), available from http://www.transitionculture.org.

14. For more information on Tradable Energy Quotas (TEQs) and its progress towards implementation in the UK see: http://www.teqs.net/. The scheme is also discussed in more depth on p.66.

15. See http://www.cityofsanctuary.org/.

"The Great Reskilling" is a Transition concept based around helping individuals and communities to learn the basic practical skills (growing food, repairing clothes, etc.) that many never learned in our cheap-energy-dependent world.

16. *Are You Happy?*, New Economics Foundation, http://tinyurl.com/25hz9q, p.35.

17. The Happy Planet Index, New Economics Foundation (nef) 2006, http://www.happyplanetindex.org/.

Chapter 7: Population and demographics

18. Three years later, in 1966, 30 world leaders signed a statement on the problem of unplanned population growth, arguing that without action there would be nearly 7 billion people by the year 2000, jeopardising people's aspirations to peaceful and happy lives: http://www.popcouncil.org/mediacenter/popstatement.html.

19. World population data from: http://www.census.gov/ipc/www/idb/worldpop.html.

A useful tool on global and national population prospects is: http://esa.un.org/unpp/.

Image source: 'So let's talk about population', Stuart Staniford, *The Oil Drum*, Dec 20th 2005, http://tinyurl.com/59k8um.

20. Wide error margins in data: 'Memorandum from David Coleman, Professor of Demography at Oxford University', *House of Commons Select Committee on Treasury Written Evidence*, Dec 2007, http://tinyurl.com/5oluxh.

21. The demographic transition is the most important concept in demography – it is a model used to explain the process of decreasing death rates followed (significantly later) by decreasing birth rates as a country changes from a pre-industrial to an industrialised economy.

For general information on this, see: http://en.wikipedia.org/wiki/Demographic_transition.

For UK-focused information on this see: 'The UK population: past, present and future', Julie Jefferies, *Focus on People and Migration: 2005*, http://tinyurl.com/6zzxc8, pp.5-6.

Further detail on life expectancy and fertility: *ibid*.

Also see 'Population Growth & Migration', *Gaia Watch UK*: http://www.gaiawatch.org.uk.

Total fertility rate defined as "the number of children that would be born to a woman if current patterns of childbearing persisted throughout her childbearing years". UK 2007 fertility figure from 'Rise in UK fertility continues', National Statistics Online, http://tinyurl.com/ywk5rg.

22. '2006-based National Population Projections', UK Statistics Authority, http://tinyurl.com/6mj7nk.

'Base population', Government Actuary's Department, http://tinyurl.com/6cfyo7.

23. Crown copyright material is reproduced with the permission of the Controller Office of Public Sector Information (OPSI). Reproduced under the terms of the click-use licence issued to the author.

Population pyramid at: http://www.statistics.gov.uk/cci/nugget.asp?id=6.

Interactive version at: http://tinyurl.com/lzrq7.

24. 'National Projections', Office for National Statistics, Oct 23rd 2007, http://www.statistics.gov.uk/cci/nugget.asp?id=1352.

This projection takes into account the change in women's state pension age from 60 to 65, which will be phased in between 2010 and 2020, and the rise in state pension age for all persons from 65 to 66 between 2024 and 2026. It is not hard to see why such changes have been considered necessary.

25. 'Population movement within the UK', Champion, T., Chapter 6 in *Focus on People and Migration*, 2005 edition. Office of National Statistics.

26. 'The Determinants of Migration Flows in England: A Review of Existing Data and Evidence', Champion *et al.* Universities of Leeds and Newcastle, http://tinyurl.com/6kh793.

27. *People and Places: A 2001 Census Atlas of the UK*, Daniel Dorling and Bethan Thomas, Policy Press, 2004. Example maps and cartograms from the book can be viewed at: http://tinyurl.com/6lmxfu.

28. 'UN: Thirty-five years to half-extinction', *The Ticker*, 24 Sept 2007, http://tinyurl.com/3czjev.

For more information on the present mass extinction see: 'Introduction to Conservation

Genetics', Richard Frankham *et al.*, Cambridge University Press, http://tinyurl.com/5lvlgw and: http://www.johnfeeney.net/.

29. One attempt to challenge this cultural story is the consideration of 'ecosystem services' – examining how much it would cost to try to replace by human and mechanical means all of the things Nature provides for us. Unsurprisingly the answer is that it would cost more than the combined value of all human goods and services globally, thus effectively demonstrating that economy is, so to speak, a subsidiary of ecology.

'The value of the world's ecosystem services and natural capital', Robert Costanza *et al.*, *Nature*, 1997, http://www.nature.com/nature/journal/v387/n6630/abs/387253a0.html.

30. *Collected Papers [by] Kenneth E. Boulding, Vol.2*, K. Boulding, Colorado Associated U. Press, 1971, p.137.

31. See for example *British Medical Journal*, July 2008 editorial, *BMJ* 2008;337:a576, http://tinyurl.com/59x2pl.

32. Guillebaud J. *Youthquake: population, fertility and environment in the 21st century*. Optimum Population Trust, 2007. http://www.optimumpopulation.org/Youthquake.pdf.

33. Population growth rates from: http://www.census.gov/ipc/www/idb/summaries.html.

34. 'UN: Thirty-five years to half-extinction', *The Ticker*, 24 Sept 2007, http://tinyurl.com/3czjev.

35. 'State pension age to rise to 68', *BBC News*, 25 May 2006: http://news.bbc.co.uk/1/hi/business/5015928.stm.

Chapter 8: Food and water

36. 'Where's the global food crisis taking us?', Ruth Gidley, *Reuters*, July 2nd 2008, http://www.alertnet.org/db/an_art/1264/2008/06/2-175818-1.htm.

37. 'UK food prices show 8.3% increase', BBC News, Sept 4th 2008, http://tinyurl.com/6fy9t2.

38. *Agriculture in the United Kingdom*, DEFRA, 2007, http://tinyurl.com/5lxbv7, Chart 7.8, p.66.

39. 'The Root of the Problem', Dr David Barling, Professor Tim Lang and Rosalind Sharpe, *RSA Journal*, Spring 2008, http://tinyurl.com/5gzfem.

40. *Food Industry Sustainability Strategy*, DEFRA, May 2006, http://tinyurl.com/6p62u6.

Food Security and the UK: An evidence and analysis paper, DEFRA, Dec 2006, http://tinyurl.com/6pax3v.

41. 'The Root of the Problem', Dr David Barling, Professor Tim Lang and Rosalind Sharpe, *RSA Journal*, Spring 2008, http://tinyurl.com/5gzfem.

Fuelling a Food Crisis, Caroline Lucas MEP, Andy Jones and Colin Hines, December 2006, http://tinyurl.com/6f9dt3, p.12.

42. *Agriculture in the United Kingdom*, DEFRA, 2007, http://tinyurl.com/5lxbv7, Chart 7.4, p.62.

43. 'The Big Question: Is changing our diet the key to resolving the global food crisis?', Jeremy Laurence, *The Independent*, Apr 16th 2008, http://tinyurl.com/55c9yx.

44. *Food Matters: Towards a Strategy for the 21st Century*, UK Government Strategy Unit, July 7th 2008, http://tinyurl.com/6g9426, p.v.

Green, Healthy and Fair: A review of government's role in supporting sustainable supermarket food, Sustainable Development Commission, Feb 2008, http://tinyurl.com/6zbzmw, pp.6-8.

45. *The Food We Waste*, WRAP, May 2008, http://tinyurl.com/5emjx2.

'Wasted food costs £10 billion: WRAP', *DEFRA website*, May 8th 2008, http://tinyurl.com/557cpl.

46. *Water for people and the environment*, Environment Agency, July 2007, http://tinyurl.com/5emnya, p.6.

47. For more information on the current and likely future impacts of climate change in the UK, see p.162.

48. *Green, Healthy and Fair: A review of government's role in supporting sustainable supermarket food*, Sustainable Development Commission, Feb 2008, http://tinyurl.com/6zbzmw, p.8.

Tim Lang interview, *Seedling*, GRAIN, July 2008, http://www.grain.org/seedling/?id=553.

49. 'Land Use and Landscape', Environment Agency website, http://tinyurl.com/6kq2c9

The state of soils in England and Wales, Environment Agency, March 2004, http://tinyurl.com/6r32qb, p.6.

Soil Association response to DEFRA discussion paper on UK food security, Sept 15th 2008, http://tinyurl.com/5v7xo8.

50. 'Organic farming and food distribution', Peter Melchett, Jan 26th 2007, http://tinyurl.com/6rq592.

'Understanding the Global Carbon Cycle', Woods Hole Research Centre, http://tinyurl.com/6alnn8.

Soil and Climate Change conference, European Commission, June 12th 2008, http://tinyurl.com/5hh9fc.

51. 'Farming', DEFRA website, http://www.defra.gov.uk/farm/index.htm.

Soil Association response to DEFRA discussion paper on UK food security, Sept 15th 2008, http://tinyurl.com/5v7xo8.

52. 'Threats of Peak Oil to the Global Food Supply', Richard Heinberg, July 2005, http://tinyurl.com/57r3wy.

Fuelling a Food Crisis, Caroline Lucas MEP, Andy Jones and Colin Hines, December 2006, http://tinyurl.com/6f9dt3, p.3.

'Joining the Dots', Chris Skrebowski, Presentation to Energy Institute Conference, London, November 10th 2004.

53. *Fuelling a Food Crisis*, Caroline Lucas MEP, Andy Jones and Colin Hines, December 2006, http://tinyurl.com/6f9dt3, p.26.

54. *Ensuring the UK's Food Security in a Changing World*, DEFRA, July 17th 2008, http://tinyurl.com/6omwwn.

Resilience in the Food Chain, Helen Peck, July 2006, http://tinyurl.com/6qf47h.

55. *Soil Association response to DEFRA discussion paper on UK food security*, Sept 15th 2008, http://tinyurl.com/5v7xo8.

Impacts of climate change on UK farming: UK Climate Impacts Programme, http://tinyurl.com/5soxb7.

'Warmer World May Mean Less Fish', United Nations Environment Programme, http://tinyurl.com/3b64pm.

Percentage of UK emissions from food chain: *Cooking up a storm: Food, greenhouse gas emissions and our changing climate*, Tara Garnett, Food Climate Research Network, Sept 2008, http://tinyurl.com/6y7x6n, p.3.

56. 'Foresight without vision', Simon Fairlie, *The Ecologist*, June 20th 2008, http://tinyurl.com/626nzr http://www.communitylandtrust.org.uk/.

57. Slow Food UK, http://www.slowfood.org.uk/.

58. "Shun meat, says UN climate chief", BBC News, Sept 7th 2008, http://tinyurl.com/6zmjuv.

59. 'The Big Question: Is changing our diet the key to resolving the global food crisis?', Jeremy Laurence, *The Independent*, Apr 16th 2008, http://tinyurl.com/55c9yx.

'An organic future?', *Living Earth*, Winter 2008 edition, p.23.

Table from: 'A method to determine land requirements relating to food consumption patterns', P.W. Gerbens-Leenes, S. Nonhebel and W. P. M. F. Ivens, *Agriculture, Ecosystems & Environment*, Volume 90, Issue 1, June 2002, pp.47-58, http://tinyurl.com/5qjhv9.

60. 'Can Britain Feed Itself?', Simon Fairlie, *The Land*, Winter 2007/8, http://tinyurl.com/2eg3qu.

61. *The Multiple Functions and Benefits of Small Farm Agriculture*, Peter Rosset, Food First/The Institute for Food and Development Policy, Sept 1999, http://tinyurl.com/6fm5m5.

62. 'Bring back the horses', Kris De Decker, Low-tech magazine, Apr 18th 2008, http://www.lowtechmagazine.com/2008/04/horses-agricult.html.

2002 Census of Agriculture, US Department of Agriculture, 2002, Volume 1, Chapter 1, Table 55, pp.58-65, http://tinyurl.com/5hnnbs.

63. 'Fifty Million Farmers', Richard Heinberg, Nov 2006, http://tinyurl.com/5hfgz4.

64. For information on Community Supported Agriculture see: http://www.soilassociation.org/csa.

65. Tim Lang interview, *Seedling*, GRAIN, July 2008, http://www.grain.org/seedling/?id=553.

UK Water Footprint: the impact of the UK's food and fibre consumption on global water resources, WWF, August 2008, http://tinyurl.com/5uyrnl.

When The Rivers Run Dry, Fred Pearce, Eden Project Books, 2007.

66. 'Food Miles professor calls for shift in food culture at Real Food debate', Julian Gairdner, *Farmers Weekly Interactive*, 25 April 2008, http://tinyurl.com/5whus5.

67. International Assessment of Agricultural Knowledge, Science and Technology for Development (IAASTD) report, Executive Summary, April 2008, http://tinyurl.com/5uprvl.

More information at: http://www.agassessment-watch.org/.

68. 'Exposed: the great GM crops myth', Geoffrey Lean, *The Independent*, Apr 20th 2008, http://tinyurl.com/3tdoed.

69. *Ibid*.

70. International Assessment of Agricultural Knowledge, Science and Technology for Development (IAASTD) report, Executive Summary, April 2008, http://tinyurl.com/5uprvl, p.4.

71. http://www.permacultureprinciples.com/principles.php.

Chapter 9: Electricity and energy

72. For further information on the coming precipitous drop-off in nuclear electricity contribution, caused by the planned decommissioning of existing plants out to 2027 see: 'Nuclear Britain', Chris Vernon, *The Oil Drum: Europe*, Jan 15th 2008, http://europe.theoildrum.com/node/3486.

73. Energy statistics taken from: *UK Energy In Brief*, Department for Business Enterprise and Regulatory Reform, July 2008, http://tinyurl.com/4onvwj and from *Meeting the Energy Challenge: A White Paper on Energy*, Department of Trade and Industry, May 2007, http://www.berr.gov.uk/whatwedo/energy/whitepaper/page39534.html.

For further information on the coming precipitous drop-off in nuclear electricity contribution, caused by the planned decommissioning of existing plants out to 2027 see: 'Nuclear Britain', Chris Vernon, *The Oil Drum: Europe*, Jan 15th 2008, http://europe.theoildrum.com/node/3486.

74. *Energy Flow Chart 2007*, DBERR, Sept 2008, http://www.berr.gov.uk/files/file46984.pdf, p.3.

75. *UK Energy In Brief*, Department for Business Enterprise and Regulatory Reform, July 2008, http://www.berr.gov.uk/whatwedo/energy/statistics/publications/in-brief/page17222.html, pp.9, 31-32.

'Fury as fuel poverty soars close to a 10-year record', Tim Webb, *The Observer*, Jan 20th 2008, http://www.guardian.co.uk/business/2008/jan/20/utilities.householdbills.

76. Department of Energy and Climate Change website: http://www.decc.gov.uk/.

77. Press Release: *RAB says UK's proposed renewable energy target is achievable*, June 18th 2008, http://tinyurl.com/5892vu.

77a. For some counter-advertising on clean coal see the 30 second video at: http://tinyurl.com/bggxpo.

78. These challenges for nuclear electricity are discussed in detail in: *The Lean Guide to Nuclear Energy: A Life-Cycle in Trouble*, David Fleming, The Lean Economy Connection: London (2007), available for download at: http://tinyurl.com/22djno.

79. *Meeting the Energy Challenge: A White Paper on Energy*, Department of Trade and Industry, May 2007, http://www.berr.gov.uk/whatwedo/energy/whitepaper/page39534.html, p.125.

80. *Grid 2.0, the next generation*, Rebecca Willis, Green Alliance (2006), http://www.green-alliance.org.uk/uploadedFiles/Publications/Grid20TheNextGeneration.pdf, p.17.

81. For full information on Tradable Energy Quotas (TEQs) and its progress towards implementation in the UK see: http://www.teqs.net/ and *Energy and the Common Purpose*, David Fleming, The Lean Economy Connection: London, 3rd edition (2007), available for download at: http://tinyurl.com/2rqwz2.

82. *Zero Carbon Britain*, Centre for Alternative Technology in collaboration with the Public Interest Research Centre, Tim Helweg-Larsen and Jamie Bull, 2007, www.zerocarbonbritain.org.

Sharon Astyk also reminds us that should power outages become a more regular

occurrence under any scenario, going without electricity is not in fact the end of the world: 'Is Electricity Really the Lifeblood of Civilisation?', Sharon Astyk, *Casaubon's Book*, June 26th 2008, http://sharonastyk.com/2008/06/26/is-electricity-really-the-lifeblood-of-civilization/.

Chapter 10: Travel and transport

83. Percentage of transport in world fuelled by oil: 'Preparing Transport for Oil Depletion' slideshow (http://tinyurl.com/4d89pt), based on *Transport Revolutions: Moving People and Freight without Oil* by Richard Gilbert and Anthony Perl, www.transportrevolutions.info.

Percentage of UK oil usage for transportation: DBERR Digest of UK Energy Statistics 2007, http://stats.berr.gov.uk/energystats/dukes07.pdf, Table 3.2, pp.84-85.

84. Transport statistics from: *Transport Statistics Great Britain*, 2007 ed., Department for Transport, http://tinyurl.com/6na7lu.

Transport Trends Great Britain, 2007 ed., Department for Transport, http://tinyurl.com/3dytuv and *Zero Carbon Britain*, Tim Helweg-Larsen and Jamie Bull, Centre for Alternative Technology in collaboration with the Public Interest Research Centre, 2007, www.zerocarbonbritain.org.

85. *Towards Transport Justice: Transport and Social Justice in an Oil-Scarce Future*, Sustrans, Ian Taylor and Lynn Sloman, April 2008, http://tinyurl.com/5kfjfn, p.4.

86. Quarter of UK households stat from: *Transport Trends Great Britain*, 2007 ed., Department for Transport, http://tinyurl.com/3dytuv, p.21.

Stats on poorest 20% from *Towards Transport Justice: Transport and Social Justice in an Oil-Scarce Future*, Sustrans, Ian Taylor and Lynn Sloman, April 2008, http://tinyurl.com/5kfjfn, p.9.

87. House of Commons Hansard written answer, 17 July 2007, http://tinyurl.com/67xkam (accessed 10/09/2008).

88. *The effect of fuel prices on motorists*, for the AA and the UK Petroleum Industry Association, Stephen Glaister and Dan Graham, Sept 2000 http://tinyurl.com/d6pt74.

89. 'Fuel theft' feature, Channel 4 News, 14 August 2008: http://tinyurl.com/5bz8dv.

90. UK Climate Impacts Programme: Transport, http://tinyurl.com/5qd4c5.

Looking Over the Horizon: Visioning and Backcasting for UK Transport Policy(VIBAT), for the Department for Transport, Robin Hickman and David Banister, Jan 2006, http://www.ucl.ac.uk/~ucft696/vibat2.html.

'Transport remains main source of health-damaging pollutants', European Environment Agency, 28 July 2008, http://tinyurl.com/5kuvdu.

91. 'The age of speed: how to reduce global fuel consumption by 75 percent', Kris De Decker, *Low-tech Magazine*, Sept 11th 2008, http://www.lowtechmagazine.com/2008/09/speed-energy.html.

92. 'Mobility behaviour', Socialdata, http://www.socialdata.de/daten/mob_e.php (note that 1,000 trips per person per year = 2.74 trips per person/day).

Focus on Personal Travel: 2005 Edition, Dept for Transport, http://tinyurl.com/8dxbeq, p.27.

93. Image Source: *Transport Revolutions: Moving People & Freight Without Oil*, Richard Gilbert and Anthony Perl, Earthscan Publications Ltd, 2007. Copyright on image owned by Ernst-Ulrich von Weizsäcker/OECD: http://tinyurl.com/8s7yyr.

94. London Freewheel: http://www.london.gov.uk/freewheel/.

Ciclovía: http://tinyurl.com/2djt8e.

World Naked Bike Ride: http://www.worldnakedbikeride.org/.

95. 'The UK Energy White Paper: An Academic Critique', Mike Pepler, *The Oil Drum: Europe*, 07 October 2007, http://europe.theoildrum.com/node/3057.

96. For more information on Tradable Energy Quotas (TEQs) and its progress towards implementation in the UK see: http://www.teqs.net/. The scheme is also discussed in more depth on p.66.

97. 'Delays ahead as road costs soar', Angela Jameson and Helen Nugent, *The Times Online*, Jan 23rd 2006, http://www.timesonline.co.uk/tol/news/article717899.ece.

98. 'In Praise of Slowness', Carl Honoré, http://www.carlhonore.com/?page_id=6.

Chapter 11: Health and medicine

99. Inefficiency of NHS buildings: 'How peak oil will affect healthcare', Stuart Jeffery, *The International Journal of Cuban Studies*, ISSN 1756-347X, http://tinyurl.com/56y343.

Taking the Temperature, New Economics Foundation, http://tinyurl.com/5rlh2p, p.2

100. 'Peak Petroleum and Public Health', Frumkin *et al.*, *Journal of the American Medical Association*, Oct 10th 2007, Vol. 298(14):1688-1690, http://jama.ama-assn.org/cgi/content/extract/298/14/1688.

101. NHS Drug budget: *Senior Manager's Guide: Drugs Bill*, The Information Centre, http://tinyurl.com/27bj8t.

102. 'It's time to face peak oil', Dan Bednarz, *Pittsburgh Post Gazette*, Dec 2nd 2007, http://tinyurl.com/5zbjgs.

103. *Material Health: A mass balance and ecological footprint analysis of the NHS in England and Wales*, Best Foot Forward Ltd, 2004.

'Transport Secretary urges hospitals to reduce reliance on the car', Department for Transport Press Notice, September 1996.

104. Sethi D, Racioppi F, Mitis F. *Youth and road safety in the European Region*. Copenhagen: WHO Regional Office for Europe, 2007.

Gill M., Goldacre M.J., Yeates D.G.R. *Changes in safety on England's roads: analysis of hospital statistics*. BMJ 2006;333:73–5.

Sethi D., Racioppi F., Baumgarten I. *et al.* Reducing *inequalities from injuries in Europe*. Lancet 2006;368:2243–50.

2005 Valuation of the Benefits of Prevention of Road Accidents and Casualties, Department for Transport, January 2007, http://tinyurl.com/5vba44.

Explanatory Memorandum To The Personal Injuries (NHS Charges) (Amounts) Regulations 2006, 2nd November 2006, http://www.opsi.gov.uk/si/si2006/draft/em/uksidem_0110752767_en.pdf.

105. Cancer and lifestyle: http://info.cancerresearchuk.org/cancerstats/causes/lifestyle/.

106. 'Health and Sustainability', *Transition Culture*, Apr 2nd 2008, http://tinyurl.com/5jsb7k.

107. UK Climate Impacts Programme: Health, http://tinyurl.com/6kwlsm.

The Health Impact of Climate Change; Promoting Sustainable Communities, Department of Health Guidance Document. April 2008. http://tinyurl.com/4p6wge.

'Climate change and human health: present and future risks', McMichael A.J., Woodruff R.E., Hales S., *The Lancet* (2006) 367:859-869, http://tinyurl.com/4fjm5w.

For more on the UK impact of climate change see p.164.

108. *Health effects of climate change in the UK, an update*, Department of Health and Health Protection Agency, 12 Feb 2008, http://tinyurl.com/ysj5as, p. iii (original 2002 report also available at this link).

109. *NHS Draft Carbon Reduction Strategy*, National Health Service Sustainable Development Unit. May 2008: http://www.sdu.nhs.uk/downloads/draft_nhs_carbon_reduction_strategy.pdf, p.3.

Also see: *The Health Impact of Climate Change: Promoting Sustainable Communities*, Department of Health, April 2008. http://tinyurl.com/4p6wge.

110. World Health Organisation, http://www.who.int/whosis/en/index.html.

111. See National Electronic Library of Infection (NeLI) site at: http://www.antibioticresistance.org.uk/.

112. *Climates and Change: The Urgent Need to Connect Health and Sustainable Development*, UK Public Health Association, 2007. http://tinyurl.com/3hy4tb.

For more information on Tradable Energy Quotas (TEQs) and its progress towards implementation in the UK see: http://www.teqs.net/. The scheme is also discussed in more depth on p.66.

113. For details of the founding One Mile Pharmacopoeia at Ruskin Mill College see: http://www.rmet.co.uk/cultural_development/cd_menu9.htm.

114. 'One in four of us will experience a mental health problem at some point in our lives', Mind, http://www.mind.org.uk/About+Mind/ (accessed Sept 9th 2008).

Chapter 12: Wildcards

115. Energy conflict as driver of war: See e.g. *Future Scenarios*, David Holmgren, 2008, http://www.futurescenarios.org/content/view/13/27/.

'Brown blunders in pledge to secure Nigeria oil', *The Independent*, 11th July 2008, http://tinyurl.com/6nwvyd.

116. 'The Ecology of Infectious Diseases', Apr 4th 2006, *University of Michigan*, http://tinyurl.com/486gfx.

117. *Climate Change and International Security: Paper from the High Representative and the European Commission to the European Council*, http://tinyurl.com/5e2skp, p.5.

Chapter 13: An overview – systems thinking

118. Rob Hopkins, *The Transition Handbook: From Oil Dependency to Local Resilience*, 1st ed. (Green Books, 2008), pp.55-56, available from http://www.transitionculture.org.

Hopkins here draws on *Resilience Thinking*, Brian Walker and David Salt, Island Press, 2006.

119. *How Nature Works: The Science of Self-Organized Criticality*, Per Bak, Springer-Verlag New York Inc. (1999 reprint).

120. *A Green New Deal: Joined-up policies to solve the triple crunch of the credit crisis, climate change and high oil prices*, Green New Deal Group, July 2008, http://www.neweconomics.org/gen/z_sys_publicationdetail.aspx?pid=258, p.4.

121. 'The Failure of Networked Systems', David Clarke (AKA aeldric), *The Oil Drum: Australia/New Zealand*, Jan 6th 2008, http://anz.theoildrum.com/node/3377.

122. For more information on Tradable Energy Quotas (TEQs) and its progress towards implementation in the UK see: http://www.teqs.net/. The scheme is also discussed in more depth on p.66.

123. A 2008 EU-commissioned report put the annual cost of forest loss (in terms of the loss of the services they provide to us) at between $2 trillion and $5 trillion every year, dwarfing even 2008's losses in the financial sector. See: 'Nature loss "dwarfs bank crisis"', Richard Black, *BBC News*, Oct 10th 2008, http://news.bbc.co.uk/1/hi/sci/tech/7662565.stm.

Original report available at: http://tinyurl.com/4aempq.

On the broader topics under discussion I recommend the article 'Herman Daly on the Credit Crisis, Financial Assets, and Real Wealth', Nate Hagens, *The Oil Drum*, Oct 13th 2008, http://www.theoildrum.com/node/4617.

124. Source: *Great Transition: The Promise and the Lure of the Times Ahead*, Paul Raskin et al., Stockholm Environment Institute, Boston, MA (2002), p.42. Available for free download at: www.tellus.org. Figure reproduced with permission.

Chapter 14: Timelines and Energy Descent Plans

125. The Kinsale EDAP can be downloaded from www.transitionculture.org/essential-info/pdf-downloads/kinsale-energy-descent-action-plan-2005/.

126. Hermione Elliott – hermione@burnout-solutions.com, www.burnoutsolutions.com.

Also see: www.imagework.co.uk.

127. *Future Scenarios*, David Holmgren (2008), www.futurescenarios.org.

Intelligent Future Infrastructure: the scenarios towards 2055, Foresight (2005), http://tinyurl.com/5sye8n.

Energy Scenarios Ireland, FEASTA (2005), www.energyscenariosireland.com.

Chapter 15: Peak oil

128. Aspen Environment Forum, March 2008, http://tinyurl.com/6d2d3s (video).

129. http://www.theoildrum.com/node/4291.

130. http://www.aspo-usa.com/fall2006/presentations/pdf/Hughes_D_OilSands_Boston_2006.pdf.

131. http://www.aspo-ireland.org/index.cfm/page/newsletter.

132. On Dec 1st 2008, Fatih Birol of the International Energy Agency publicly discussed the IEA's *World Energy Outlook 2008*, which includes the first public field-by-field analysis of the world's 800 largest oil fields (which together make up more than three quarters of global reserves and more than two-thirds of global oil production). Birol announced that just to compensate for decline and keep oil production flat over the next twenty-five years, we would need to generate 45 million barrels/day in *new production*. This is roughly the equivalent of generating four new Saudi Arabias (the biggest oil exporter in the world), and does not even allow for any growth in demand.

Video of full presentation available at: http://www.theoildrum.com/node/4952 (these comments at 13m30 in).

133. For further discussion on this point see: 'Geopolitical Disruptions #1: Theory of Disruptions to Oil & Resource Supply', Jeff Vail, *The Oil Drum*, Aug 14th 2008: http://www.theoildrum.com/node/4373.

133a. Figures taken from International Energy Agency (IEA) *World Energy Outlook 2008*: http://www.worldenergyoutlook.org/2008.asp

and US Dept. of Energy/Energy Information Administration (EIA) *International Energy Outlook*.

134. *The Times*, Apr 8th 2006, http://tinyurl.com/2slp5a.

135. *The Financial Times*, Oct 31st 2007, http://tinyurl.com/2d46x6.

136. *Energy Bulletin*, http://www.energybulletin.net/37000.html.

137. *Sydney Morning Herald*, Jan 15th 2008, http://tinyurl.com/ypjcaw.

138. http://www.whitehouse.gov/news/releases/2007/11/20071107-5.html.

139. *Reuters*, Mar 17th 2008 http://www.reuters.com/article/politicsNews/idUSLAW00006020080317.

140. Graph source: 'Olduvai Revisited 2008', Luis de Sousa, *The Oil Drum: Europe*, Feb 28th 2008, http://europe.theoildrum.com/node/3565.

Conversion rate: 1 barrel of oil = 0.146 tons of oil, 6.841 barrels of oil = 1 ton of oil.

141. Source: www.eroei.com.

142. 'A framework for analysing alternative energy: Net energy, Liebig's Law and Multicriteria Analysis', Nathan Hagens and Kenneth Mulder, *Renewable Energy Systems: Environmental and Energetic Issues*, edited by David Pimentel. Reprinted with permission from Nathan Hagens.

143. The overlooked importance of EROEI can be seen in the fact that 'total liquids' production is simply measured in 'barrels per day', yet a barrel of ethanol has only roughly two thirds the energy content of a barrel of crude oil. A barrel of Liquefied Natural Gas (LNG) has only roughly three quarters the energy content of a barrel of crude oil. Source: http://europe.theoildrum.com/node/4007.

144. It should be noted that there are many different oil prices, as there are many different varieties and grades of crude oil, but Brent blend crude oil from the North Sea sold at London's International Petroleum Exchange is generally considered the 'benchmark' for the global oil price.

The latest Brent oil price can be accessed on the BBC News website at: http://tinyurl.com/ov5l5.

145. It is worth noting that the prices of other energy resources (such as natural gas) tend to change broadly in keeping with the oil price.

For further details on the 1970s oil crises see '1973 Oil Crisis', *Wikipedia*, http://tinyurl.com/8llqr and '1979 Oil Crisis', *Wikipedia*, http://tinyurl.com/nkjoz.

146. Image by TomTheHand, licensed under the Creative Commons Attribution ShareAlike 3.0 Unported License. Original (including prices from 1861 and explanation for omission of 2008 data) available at: http://tinyurl.com/5kc865.

147. This point is explored in more depth in: 'Countdown to $200 oil (10) – oil at $115!!', Jerome a Paris, *The Oil Drum: Europe*, Aug 10th 2008, http://europe.theoildrum.com/node/4399.

148. Peaking of World Oil Production: Impacts, Mitigation, and Risk Management, Robert Hirsch et al., Feb 2005, http://tinyurl.com/gqwge.

149. *The Great Emissions Rights Give-Away*, FEASTA/nef, Mar 2006, http://www.feasta.org/documents/energy/emissions2006.pdf.

150. *Ibid*.

151. 'Peaking of World Oil Production: Impacts, Mitigation, and Risk Management', Robert Hirsch et al., Feb 2005, http://tinyurl.com/gqwge.

152. Statistics on goods and farming from 'What they don't want you to know about the coming oil crisis', Jeremy Leggett, *The Independent*, Jan 20th 2006, http://tinyurl.com/6f8r9n.

Statistic on food from Professor Tim Lang presentation: *Food Security: are we sleep-walking into a crisis*, Mar 4th 2008, City University London.

More on food and oil can be found on p.53.

153. 'Sustainable Bioenergy: A Framework for Decision Makers', UN-Energy, http://www.fao.org/docrep/010/a1094e/a1094e00.htm, p.39.

154. For information see: http://energyshortage.org/, http://www.energybulletin.net/.

Chapter 16: Peak oil and climate change – the interplay

155. Commenting on a draft of this section David Fleming of The Lean Economy Connection highlighted that later fossil-fuel supply limits would only be a benefit to peak oil adaptation if we used the additional time and resources to prepare. An alternative possibility is what he terms the "Tragedy of the Breathing Space" – that we would use the opportunity to create yet more complex fossil-fuel-dependent systems and so make the eventual adaptation even harder. This is true and important, but we are here searching for our best available course of action.

156. The President of Ecuador offered in April 2007 to begin this process, leaving the oil in Yasuni National Park undisturbed to protect the park's biodiversity and indigenous peoples, but only if the international community will compensate the country with half of the forecasted lost revenues. For details see: 'Ecuador Seeks Compensation to Leave Amazon Oil Undisturbed', *Environment News* Service, Apr 24th 2007, http://tinyurl.com/2tdsse.

157. My thanks to Chris Vernon of *The Oil Drum: Europe* for this insight. The basis for this claim is outlined in Appendix A.

See p.121 for an explanation of the Energy Return On Energy Invested (EROEI) concept.

Chapter 17: Climate change explained

158. Further discussion of the IPCC's strengths and shortcomings will follow in the next chapter.

159. There is also a warming effect from the geothermal energy at the Earth's core, but this is sufficiently small and stable that for our purposes we can ignore it.

160. Parts per million is the ratio of the number of greenhouse gas molecules to the total number of molecules of dry air. For example, 300ppm means 300 molecules of a greenhouse gas per million molecules of dry air. Strictly speaking, concentrations are measured in *parts*

per million by volume (ppmv), but this is widely abbreviated to ppm. Don't be confused if some papers refer to ppmv.

161. The distance between each line and the next in these illustrative graphs is explained as follows. Emissions are not the sole determinant of atmospheric greenhouse gas concentrations due to the Earth's natural 'carbon sinks' (see p.142) which soak up some of our emissions. Concentrations are not the sole determinants of radiative forcing due to other forcings which will be discussed in the next few pages. The time delay between radiative forcing and temperature increase is caused by the thermal inertia of the planet – it has great mass and therefore takes time to warm or cool. It takes around 25 years for the first half of the temperature effect to manifest, with the next quarter taking around 150 years and the last quarter many centuries.

162. These illustrative graphs do not include the effects of climate feedbacks such as carbon sink degradation. These effects are discussed from p.142 onwards. Also see: http://tinyurl.com/6y3uqc.

163. Figures from IPCC 2007: http://ipcc-wg1.ucar.edu/wg1/Report/AR4WG1_Print_Ch02.pdf, Table 2.14, p.212.

More detail on GWP available at: http://en.wikipedia.org/wiki/Global_warming_potential – note that the GWP for a mixture of gases cannot be determined from the GWP of the constituent gases by any form of simple linear addition.

164. There is also a separate but related concept called *Carbon Dioxide equivalent*. This gives the amount of CO_2 that would have the same GWP as a given amount of a given gas (or mixture of gases). It is simply calculated by multiplying the GWP of the gas by the given amount (mass) of gas. For example, over a 100-year period methane has a GWP of 25, so 1 gram of methane has a *Carbon Dioxide equivalent* value of 25 grams.

In practice, since *Carbon Dioxide equivalent* is expressed as a mass (grams, tonnes etc.), and *Equivalent Carbon Dioxide (CO2e)* is expressed as a concentration (usually in parts per million), they are not easily confused, despite the similar names.

165. The IPCC considered the so-called 'Kyoto basket' of greenhouse gases (GHGs). Under the Kyoto Protocol, signatories committed to control emissions of a 'basket' of six GHGs – carbon dioxide, methane, nitrous oxide, HFCs, PFCs and SF6.

The IPCC calculation of CO2e(Kyoto) is detailed by Gavin Schmidt of NASA at: http://www.realclimate.org/index.php/archives/2007/10/CO2-equivalents/.

166. These negative forcings include the so-called 'global dimming' effect.

For more on this crucial consideration see: "On avoiding dangerous anthropogenic interference with the climate system: formidable challenges ahead", V. Ramanathan and Y. Feng, *Proceedings of the National Academy of Sciences*, vol. 105, 23 September 2008, pp. 14245-14250, http://tinyurl.com/cqylaz.

IPCC CO2e(Total) figure: http://www.ipcc.ch/pdf/assessment-report/ar4/syr/ar4_syr.pdf, Notes to table 5.1, p.67.

167. IPCC 2005 CO_2 levels: http://www.ipcc.ch/pdf/assessment-report/ar4/syr/ar4_syr_spm.pdf.

168. IPCC 2001 figures: http://www.grida.no/climate/ipcc_tar/wg1/248.htm, Table 6.7.

1995/2007: http://ipcc-wg1.ucar.edu/wg1/Report/AR4WG1_Print_Ch02.pdf, Table 2.14, p.212.

169. Error ranges: http://www.ipcc.ch/pdf/assessment-report/ar4/wg3/ar4-wg3-chapter1.pdf, p.102.

170. Emissions rates: http://www.ipcc.ch/pdf/assessment-report/ar4/wg3/ar4-wg3-spm.pdf, p.3.

Up-to-date measurements of atmospheric CO_2 concentrations are always subject to revisions, pending recalibrations of reference gases and other quality control checks. Trends and 2008 figure taken from: http://www.esrl.noaa.gov/gmd/ccgg/trends/ (site accessed August 2008).

Pre-industrial CO_2 levels from: http://www.noaanews.noaa.gov/stories2005/s2412.htm.

Chapter 18: Climate change – the IPCC

171. The next IPCC Assessment Report (AR5) will be finalised in 2014, with a target date of early 2013 for the release of its key Working Group 1 Report. It is noteworthy that this falls *after* the "defining moment" described by IPCC Chairman Rajendra Pachauri.

It is perhaps worth mentioning climate denialists at this point. A collection of short clear answers debunking the common contrarian arguments is maintained at: http://gristmill.grist.org/skeptics.

172. IPCC Special Report on Emissions Scenarios (SRES), Summary for Policy Makers, p.6, http://www.ipcc.ch/pdf/special-reports/spm/sres-en.pdf.

SRES assumptions on resource availability are discussed in depth at: http://www.ipcc.ch/ipccreports/sres/emission/104.htm.

173. Global fossil fuel emissions from Carbon Dioxide Information Analysis Centre: http://cdiac.ornl.gov/trends/emis/tre_glob.html.

174. Kharecha and Hansen's paper is available at: http://arxiv.org/abs/0704.2782, with further discussion of it at: http://www.theoildrum.com/node/2559 Bear in mind that one tonne of carbon = 3.644 tonnes CO_2.

Dave Rutledge of the California Institute of Technology looks at more realistic fossil fuel reserves and their implications for the IPCC scenarios in a very clear presentation here: http://rutledge.caltech.edu/.

175. 'Implications of peak oil for atmospheric CO₂ and climate', Drs Pushker Kharecha and James Hansen, NASA, http://arxiv.org/pdf/0704.2782v3, p.3.

176. Also see Appendix A.

On Amazon rainforest drying and carbon release see: 'A disaster to take everyone's breath away', Geoffrey Lean, *The New Zealand Herald*, http://tinyurl.com/2gjeqd.

Methane hydrates are ice crystal formations containing significant quantities of methane, found in both deep sedimentary deposits and on the ocean floor. There has been some excitement over the potential of methane hydrates as fuel, see e.g.: 'Japan's Arctic methane hydrate haul raises environment fears', Leo Lewis, *The Times Online*, Apr 14th 2008, http://tinyurl.com/4e9exo.

For Jean Laherrère's technical discussion of methane hydrates see: 'Hydrates updated', Jean Laherrère, *The Oil Drum: Europe*, Apr 17th 2008, http://europe.theoildrum.com/node/3819.

See p.121 for an explanation of the Energy Return On Energy Invested (EROEI) concept.

177. The B1 scenario (as with the others) makes a number of assumptions about the future storyline of the human world. These are detailed at: http://www.ipcc.ch/ipccreports/sres/emission/094.htm#1.

However, these are not directly relevant here. We have different reasons for believing this to be the IPCC's most realistic future emissions trajectory.

It is important to note that emissions are at present growing faster than projected under any of the IPCC scenarios, but the limits to available fuels will ensure that this trend cannot continue over the next twenty years (the scope of this book).

178. Values in Gt CO₂ shown in labels are cumulative emissions out to 2100 based on values in Table 5-2, Section 5, IPCC Special Report on Emissions Scenarios (SRES): http://tinyurl.com/55p3wg.

The SRES scenarios shown are: the marker scenario for B1 (IMAGE model); the illustrative scenario for A1FI (MiniCAM model); and the AIB marker scenario (AIM model). Source: SRES Final Data (version 1.1, July 2000) – http://sres.ciesin.columbia.edu/final_data.html.

179. From IPCC Press Release, 18th Sept 2007 – http://tinyurl.com/2chmus.

Full report: *Climate Change 2007: Impacts, Adaptation and Vulnerability. Contribution of Working Group II to the Fourth Assessment Report of the Intergovernmental Panel on Climate Change*, M.L. Parry, O.F. Canziani, J.P. Palutikof, P.J. van der Linden and C.E. Hanson, Eds., Cambridge University Press, Cambridge, UK, 7-22. http://www.ipcc-wg2.org/index.html.

180. Remember that it takes around 25 years for the first half of the temperature effect to manifest, with the next quarter taking around 150 years and the last quarter many centuries.

181. *Contribution of Working Group II to the Fourth Assessment Report of the Intergovernmental Panel on Climate Change, Summary for Policymakers*, http://www.ipcc.ch/pdf/assessment-report/ar4/wg2/ar4-wg2-spm.pdf, p.14.

182. *Ibid.*, p.8.

183. 'Climate Change and International Security: Paper from the High Representative and the European Commission to the European Council', http://tinyurl.com/5e2skp, p.1.

184. IPCC 2007 Synthesis report: http://www.ipcc.ch/pdf/assessment-report/ar4/syr/ar4_syr.pdf, p.54.

184a. For a more detailed investigation of the implications of different levels of global temperature rise see: *Six Degrees: Our Future on a Hotter Planet*, Mark Lynas, National Geographic Society (2008).

185. Rahmstorf S., Cazenave A., Church J., Hansen J., Keeling R., Parker D., Somerville R., (2007) Recent climate observations compared to projections. *Science*, v316. Published online 1 February 2007; 10.1126/science.1136843. Reprinted with permission from AAAS. http://tinyurl.com/5t79fd.

186. IPCC 2007 Synthesis report: http://www.ipcc.ch/pdf/assessment-report/ar4/syr/ar4_syr.pdf, Notes to table 5.1, p.67.

187. For a summary of all climate feedback dynamics see: *The Westminster Briefing: An Introduction to Climate Dynamics*, David Wasdell, http://tinyurl.com/5te8ds.

Climate Change and Trace Gases, Hansen *et al.* (2007), Philosophical Transactions of the Royal Society, http://www.planetwork.net/climate/Hansen2007.pdf.

188. Examples from: 'Beck to the future', *RealClimate*, May 1st 2007, http://tinyurl.com/5cjyhb.

General information – David Wasdell: http://www.bigpicture.tv/videos/watch/371bce7dc (video).

189. Paper giving 18% figure: 'Contributions to accelerating atmospheric CO₂ growth from economic activity, carbon intensity, and efficiency of natural sinks', Canadell et al., PNAS, http://www.pnas.org/content/104/47/18866.full.pdf+html.

Quote from press release: 'Decline in uptake of carbon emissions confirmed', *CSIRO*, http://www.csiro.au/news/CarbonEmissionsConfirmed.html.

190. 'Rapid rise in global warming is forecast', Lewis Smith, *The Times Online*, May 18th 2007, http://www.timesonline.co.uk/tol/news/uk/science/article1805870.ece.

191. For information on second order feedbacks see 'The Westminster Briefing:

Accelerated Climate Change and the Task of Stabilisation', David Wasdell, http://www.apollo-gaia.org/Presentation5.pdf.

192. *Global warming: East-West connections*, Hansen and Sato (2007), www.columbia.edu/~jeh1/2007/EastWest_20070925.pdf, pp.7-10.

For further explanation and discussion of 'climate sensitivity' see: 'Target CO2', Gavin Schmidt, Real Climate (blog), Apr 7 2008, http://tinyurl.com/dc5mgk

and: *Climate Safety*, Richard Hawkins et al., First Edition (Public Interest Research Centre, 2008), pp. 15-17. Available for download at: http://www.climatesafety.org/.

193. *IPCC AR4 WG1*, http://www.ipcc.ch/pdf/assessment-report/ar4/wg1/ar4-wg1-chapter7.pdf, p.566.

194. 'The IPCC: As good as it gets', Professor Martin Parry, *BBC News*, Nov 13th 2007: http://news.bbc.co.uk/2/hi/science/nature/7082088.stm.

There is also controversy over the change in IPCC Chair in 2002, as Dr Robert Watson was replaced just over a year after Exxon sent the following memo to the White House asking 'Can Watson Be Replaced Now At The Request of the US?' http://www.nrdc.org/media/docs/020403.pdf.

More details at: http://www.newscientist.com/article.ns?id=dn2191.

195. 'How climate change will affect the world', David Adam, *The Guardian*, Sept 19th 2007, http://www.guardian.co.uk/environment/2007/sep/19/climatechange.

196. 'The IPCC: As good as it gets', Professor Martin Parry, *BBC News*, Nov 13th 2007: http://news.bbc.co.uk/2/hi/science/nature/7082088.stm.

197. The draft of the 2007 *Summary for Policymakers* with the changes highlighted can be found here: http://tinyurl.com/67c2o3.

Peter Wadhams quote from: 'Climate report was "watered down"', *New Scientist*, March 10th 2007, http://environment.newscientist.com/article/mg19325943.900.

Further detail and discussion in: 'Political Corruption of the IPCC Report?', David Wasdell, Feb 16th 2007, http://www.meridian.org.uk/_PDFs/IPCC.pdf.

Chapter 19: Climate change – a reality check

198. *Climate Code Red: The Case for a Sustainability Emergency*, David Spratt and Philip Sutton, Friends of the Earth Australia (February 2008), http://tinyurl.com/65hk7b, p.24.

199. *Global warming: East-West connections*, Hansen and Sato (2007), www.columbia.edu/~jeh1/2007/EastWest_20070925.pdf, p.13.

200. *Climate Code Red: The Case for a Sustainability Emergency*, David Spratt and Philip Sutton, Friends of the Earth Australia (February 2008), http://tinyurl.com/65hk7b, p.40.

Hansen and Sato ref: *Climate Change and Trace Gases*, Hansen et al. (2007), Philosophical Transactions of the Royal Society, http://www.planetwork.net/climate/Hansen2007.pdf.

201. *Target atmospheric CO2: Where should humanity aim?*, Hansen et al. (2008), http://arxiv.org/abs/0804.1126, p.1

202. *The United Nations Framework Convention on Climate Change*, Article 2. http://unfccc.int/essential_background/convention/background/items/1353.php.

Holdren quote from: *Planet Earth: We Have a Problem*, All Party Parliamentary Climate Change Group in association with the Meridian Programme, p.125.

203. *Climate Code Red: The Case for a Sustainability Emergency*, David Spratt and Philip Sutton, Friends of the Earth Australia (February 2008), http://tinyurl.com/65hk7b, p.19.

204. *Ibid*, p.22.

In the previous chapter we considered the IPCC figures on warming due from current CO2 concentrations. A 2008 paper in the *Proceedings of the National Academy of Sciences* examined the impacts of air pollution (which blocks sunlight and thus reduces temperatures – the effect known as 'global dimming') and found that this is masking the full extent of the warming effect from greenhouse gas concentrations. Building on the IPCC's work, the paper finds that if air pollution reduces – as it is expected to do – then 2005 atmospheric concentrations could commit us to around 2.4 degrees of warming above pre-industrial temperatures, with about 90% of this warming taking place this century. See: "On avoiding dangerous anthropogenic interference with the climate system: Formidable challenges ahead", V. Ramanathan and Y. Feng, *Proceedings of the National Academy of Sciences*, vol. 105, 23 September 2008, pp. 14245-14250, http://tinyurl.com/cqylaz.

205. *Climate Code Red: The Case for a Sustainability Emergency*, David Spratt and Philip Sutton, Friends of the Earth Australia (February 2008), http://tinyurl.com/65hk7b, p.27.

206. Up-to-date measurements of atmospheric CO2 concentrations are always subject to revisions, pending recalibrations of reference gases and other quality control checks. 385ppm figure taken from http://www.esrl.noaa.gov/gmd/ccgg/trends/ (site accessed August 2008).

207. The campaigning organisation 350.org (http://350.org/) are taking this scientific advice seriously and spreading the message that 350 (ppm) is the number we all need to know and demand, with the aim of shifting the international negotiations towards scientific realism. As discussed on p.142, climate feedback mechanisms mean that low atmospheric CO2 concentrations do not automatically lead to declining radiative

forcing, but this is nonetheless a valuable campaign.

208. 'Climate Change and Trace Gases', Hansen et al.(2007), *Philosophical Transactions of the Royal Society*, http://www.planetwork.net/climate/Hansen2007.pdf.

209. For an explanation of radiative forcing, see p.130.

Techno-fixes, a critical guide to climate change technologies, Claire Fauset (2008), Corporate Watch, http://www.corporatewatch.org.

'Energy and Material Balance of CO_2 Capture from Ambient Air', Frank Zeman, *Environmental Science & Technology*, Sept 26th 2007., Vol. 41, No. 21, pp7558-7563. 10.1021/es070874m, http://tinyurl.com/5eujjv.

210. 'Comment on 'Modern-age buildup of CO_2 and its effects on seawater acidity and salinity' by Hugo A. Loáiciga', Caldeira, K., D. Archer *et al.* (2007) *Geophysical Research Letters* 38: L18608, http://tinyurl.com/6lm7jb.

NASA: 'Oceans may soon be more corrosive than when the dinosaurs died', NASA media release, 20th Feb 2006, http://earthobservatory.nasa.gov/Newsroom/MediaAlerts/2006/2006022021812.html.

Eilperin, J., 'Growing acidity of oceans may kill corals', *Washington Post*, July 5th 2006, www.washingtonpost.com/wp-dyn/content/article/2006/07/04/AR2006070400772.html.

211. See: http://www.virginearth.com/.

212. 'Understanding the Global Carbon Cycle', *Woods Hole Research Centre*, http://tinyurl.com/6alnn8.

Prof. Rattan Lal, Ohio University Carbon Management and Sequestration Centre, Soil and Climate Change conference, European Commission, June 12th 2008, http://tinyurl.com/5hh9fc.

213. *Priority One*, Allan J. Yeomans, Biosphere Media 2007, http://tinyurl.com/3224od, pp.89-91.

Prof. Rattan Lal, Ohio University Carbon Management and Sequestration Centre, Soil and Climate Change conference, European Commission, June 12th 2008, http://tinyurl.com/5hh9fc.

'Organic farming and food distribution', Peter Melchett, Jan 26th 2007, http://tinyurl.com/6rq592.

214. *Energy and the Common Purpose*, David Fleming, 3rd edition, http://www.teqs.net/book/teqs.pdf, p.39.

215. Full article recommended: 'The Greatest Danger', Joanna Macy, *YES! Magazine*, Spring 2008, http://www.yesmagazine.org/article.asp?ID=2295.

216. A startling example of this being the fungi that have developed to *eat* and directly live off the very radiation at Chernobyl. See: 'Eating Radiation: A New Form of Energy', David Ewing Duncan, *Technology Review*, May 29th 2007, http://www.technologyreview.com/blog/duncan/17611/.

Chapter 20: Peak oil in the UK

217. *Meeting the Energy Challenge: A White Paper on Energy*, Department of Trade and Industry, May 2007, http://www.berr.gov.uk/whatwedo/energy/whitepaper/page39534.html, p.6.

218. For full details see: 'UK Government Response to Peak Oil Petition', Chris Vernon, *The Oil Drum: Europe*, Oct 3rd 2007, http://www.theoildrum.com/node/3045.

219. *World Energy Outlook 2004 Edition*, http://www.worldenergyoutlook.org/2004.asp, p.47.

220. *World Energy Outlook 2004 Edition*, http://www.worldenergyoutlook.org/2004.asp, p.122.

221. 'IEA says oil prices will stay "very high", threatening global growth', James Kanter, *International Herald Tribune*, Oct 31st 2007, http://www.iht.com/articles/2007/10/31/business/oil.php.

222. 'IEA: without Iraqi Oil, we'll be in deep trouble by 2015', Jerome a Paris, *The Oil Drum: Europe*, June 29th 2007, http://europe.theoildrum.com/node/2721.

223. 'We can't cling to crude: we should leave oil before it leaves us', Fatih Birol, *The Independent*, March 2nd 2008, http://tinyurl.com/34qexo.

224. Internationale Politik, April 2008, http://tinyurl.com/4quc6z (translated from the German).

225. *World Energy Outlook 2008*, International Energy Agency, November 2008, http://www.worldenergyoutlook.org/2008.asp, Executive Summary, p.3.

226. *Communication on BERR Fossil Fuel Price Assumptions*, Department for Business Enterprise & Regulatory Reform, May 2008, http://www.berr.gov.uk/files/file46071.pdf.

227. 'We must all act together', Gordon Brown, *The Guardian*, May 28th 2008, http://tinyurl.com/4lbnm7.

228. 'Speech to the Union for the Mediterranean Summit', Gordon Brown, July 13th 2008, http://www.number10.gov.uk/Page16313.

229. From Euan Mearns' essential article: 'A State of Emergency' on *The Oil Drum: Europe*, June 25th 2008, http://europe.theoildrum.com/node/4188.

230. *Meeting the Energy Challenge: A White Paper on Energy*, Department of Trade and Industry, May 2007, http://www.berr.gov.uk/whatwedo/energy/whitepaper/page39534.html, p.106.

UK Energy In Brief, Department for Business Enterprise and Regulatory Reform, July 2008, http://www.berr.gov.uk/whatwedo/energy/statistics/publications/in-brief/page17222.html, p.17.

Financial figures from 'A State of Emergency', Euan Mearns, *The Oil Drum: Europe*, June 25th 2008, http://europe.theoildrum.com/node/4188. These figures are clearly dependent on the future price of oil, which is inherently unpredictable. The underlying assumption here is an oil price growing at 25% per year from the $110 per barrel that was the 2008 average when the article was published. Even if substantially lower future prices are postulated, the trend remains deeply concerning.

231. *Meeting the Energy Challenge: A White Paper on Energy*, Department of Trade and Industry, May 2007, http://www.berr.gov.uk/whatwedo/energy/whitepaper/page39534.html, p.109.

232. Global peak exports: 'Is a Net Oil Export Hurricane Hitting the US Gulf Coast?', Jeffrey J. Brown, *The Oil Drum*, June 2nd 2008, http://www.theoildrum.com/node/4092.

233. 'Why UK Natural Gas Prices Will Move North of 100p/Therm This Winter', Rune Likvern, *The Oil Drum: Europe*, June 24th 2008, http://europe.theoildrum.com/node/4193.

UK Energy In Brief, Department for Business Enterprise and Regulatory Reform, July 2008, http://www.berr.gov.uk/whatwedo/energy/statistics/publications/in-brief/page17222.html, p.21.

234. 'UK Energy Security', Euan Mearns, *The Oil Drum: Europe*, Oct 25th 2007, http://europe.theoildrum.com/node/3130.

235. 'Energy firms to raise bills yet again', Tim Webb, *The Observer*, Apr 20th 2008, http://www.guardian.co.uk/business/2008/apr/20/oil.householdbills.

236. Liquefied Natural Gas (LNG) is natural gas which has been liquefied by reducing its temperature to -160°C at atmospheric pressure, usually to allow for transportation by ship.

237. 'UK Natural Gas Prices, Already At Historically High Levels, Set To Rise', Doug Low, *The Oil Drum: Europe*, Mar 18th 2008, http://europe.theoildrum.com/node/3751.

238. For more information on the Tradable Energy Quotas (TEQs) idea and its progress towards implementation in the UK see: http://www.teqs.net/. The scheme is also discussed in more depth on p.66.

239. In March 2006 the National Grid announced its first ever Gas Balancing Alert, which is defined to be a genuine indication of natural gas shortage and to signal that demand reduction is required to avert an emergency. The immediate triggers were a cold snap in the weather and a fire at a gas storage site, but the medium-term picture suggests it won't be long before we see another. http://www.nationalgrid.com/uk/Gas/OperationalInfo/GBA/, http://www.ageconcern.org.uk/AgeConcern/ftf_winter_deaths.asp.

Chapter 21: Climate change in the UK

240. http://www.defra.gov.uk/environment/climatechange/uk/progress/index.htm These figures exclude international travel and transport emissions, and since 2006 have been based purely on the number of emissions permits issued by our Government under the European Union Emissions Trading Scheme, despite the fact that UK companies buy in permits issued by other governments. See: 'UK in "delusion" over emissions', Roger Harrabin, *BBC News*, July 31st 2008, http://tinyurl.com/6q8rvo.

241. '2005 UK climate change sustainable development indicator and greenhouse gas emissions final figures'. Crown copyright material is reproduced with the permission of the Controller Office of Public Sector Information (OPSI). Reproduced under the terms of the click-use licence issued to the author.

242. 'Government set to miss carbon cut target', Goska Romanowicz, *Environmental Data Interactive Exchange*, March 28th 2006, http://tinyurl.com/7r47ob.

243. 'On creating a low carbon economy', June 26th 2008, http://www.number10.gov.uk/Page16141.

244. The full text of the Climate Change Act 2008 can be found here: http://tinyurl.com/99qcmq.

245. Environmental Audit Committee Ninth Report – http://www.publications.parliament.uk/pa/cm200607/cmselect/cmenvaud/740/74010.htm.

246. 'Audit of UK Greenhouse Gas emissions to 2020: will current Government policies achieve significant reductions?', Maslin *et al.*, *UCL Environment Institute*, http://tinyurl.com/68n6cn, p.2.

Also see: 'Never mind nukes, we're missing our carbon target', Chris Goodall, *The Independent on Sunday*, May 27th 2007, http://news.independent.co.uk/business/comment/article2586535.ece.

247. http://www.hm-treasury.gov.uk/media/3/2/Summary_of_Conclusions.pdf, p. vii.

Full Stern Review final report available at: http://tinyurl.com/bfenxp.

Stern also published another short report on April 30th 2008: http://tinyurl.com/6g4kzu.

248. A footnote in Chapter 13 (p.296) of the full Review does explain that: "The world is already at around 430ppm CO_2e if only the greenhouse gases covered by the Kyoto Protocol are included; but aerosols reduce current radiative forcing", but this is clearly likely to reach a more limited audience than the summary.

George Monbiot's comments on the inhumane nature of Stern's calculations are also worth noting: http://tinyurl.com/bte7o7.

249. Source: IPCC: http://www.ipcc.ch/pdf/assessment-report/ar4/wg3/ar4-wg3-spm.pdf, p.15.

250. UN Environment Programme Press Release, Dec 2007: http://tinyurl.com/55shg7.

251. For more information on Tradable Energy Quotas (TEQs) and its progress towards implementation in the UK see: http://www.teqs.net/. The scheme is also discussed in more depth on p.66.

252. UK Climate Impacts Programme: http://www.ukcip.org.uk/.

The work of the UK Phenology Network may also be of interest: http://www.naturescalendar.org.uk/.

253. A very unusual image – a virtually cloudless UK – was captured from space during this heatwave and can be viewed at: http://tinyurl.com/a85hze.

254. 'A new daily Central England Temperature Series', D.E. Parker, T.P. Legg, and C.K. Folland. 1992, 1772-1991. *Int. J. Clim.*, Vol 12, p317-342.

255. 'The 2007 Eastern US Spring Freeze: Increased Cold Damage in a Warming World', Lianhong Gu *et al.*, *BioScience*, vol. 58, p.253, http://tinyurl.com/5fpvlx.

256. 'The Climate of the UK and recent trends', Geoff Jenkins *et al.*, *UK Climate Impacts Programme*, Dec 2007, http://www.ukcip.org.uk/index.php?option=com_content&task=view&id=469&Itemid=477.

'UK Marine Climate Change Impacts', *Marine Climate Change Impacts Partnership*, 2008, http://www.mccip.org.uk/arc/2007/default.htm.

257. An excellent explanation of the relationship between weather and climate, and the basics of climate prediction, can be found here: http://www.begbroke.ox.ac.uk/climate/interface.html

258. *UK Climate Impacts Programme 2002, Headline Messages*, http://tinyurl.com/5d43mz.

Closing thoughts

259. 'The Greatest Danger', Joanna Macy, *YES! Magazine*, Spring 2008, http://www.yesmagazine.org/article.asp?ID=2295.

Appendix A: Substitution problem calculation

260. See p.121 for an explanation of the Energy Return On Energy Invested (EROEI) concept.

261. 'Unconventional Crude', Elizabeth Kolbert, *The New Yorker*, Nov 12th 2007, http://tinyurl.com/5kwdha.

262. Current production levels are around 2.3 million b/day.

Ray Leonard is Vice-President-Eurasia with the Kuwait Energy Company. He was formerly Amoco's Director of New Ventures for the Soviet Union, Eastern Europe and China, Exploration VP for a newly formed company in Kazakhstan and VP-Exploration and New Ventures for Yukos. He holds a B.S. in Geology from the University of Arizona and an M.A. in Geology from the University of Texas-Austin. His analysis is summarised at: http://tinyurl.com/6bba7y. Slides at: http://tinyurl.com/6pjqld.

David Hughes is a geologist with 35 years experience studying the energy resources of Canada for the Geological Survey of Canada and the private sector, and is the leader of their National Coal Inventory. He is also 'Team Leader for Unconventional Gas' for the 'Canadian Gas Potential Committee', an organisation which publishes Canada's most authoritative assessments of national natural gas potential. His analysis is available at: http://tinyurl.com/yefrgv.

263. Robert Hirsch's *Peaking of World Oil Production: Impacts, Mitigation, and Risk Management*, http://tinyurl.com/gqwge, p.79. It should also be noted that Hirsch's assumption of essentially unlimited high-quality coal supplies for CTL is questionable (quite apart from the climate impacts it would imply). For detailed information on 'peak coal' see: 'Coal: The Roundup', Chris Vernon, The Oil Drum: Europe, July 12th 2007, http://tinyurl.com/6mlof9.

264. 'Fermenting the Food Supply – Revisited', Stuart Staniford, *The Oil Drum*, Aug 16th 2008, http://www.theoildrum.com/node/4422 Also 'Global Ethanol and Biodiesel Outlook 2008-2015', *Biofuels Digest*, July 2nd 2008, http://tinyurl.com/5fnnee.

265. Even without considering the possible difficulties in sourcing financing for large-scale energy projects in a post-peak world.

266. Amazon rainforest drying and carbon release: 'A disaster to take everyone's breath away', Geoffrey Lean, *The New Zealand Herald*, http://tinyurl.com/2gjeqd.

Appendix B: The Transition Timeline's relationship with Zero Carbon Britain

267. *Zero Carbon Britain*, Centre for Alternative Technology in collaboration with the Public Interest Research Centre, Tim Helweg-Larsen and Jamie Bull, 2007, www.zerocarbonbritain.org.

Index

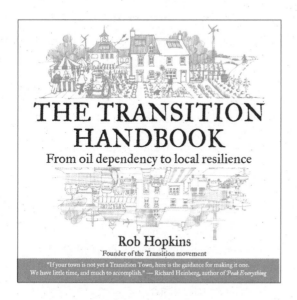

"This book by the visionary architect of the Transition movement is a must-read, labelled 'immediate'. Growing numbers with their microscopes trained on peak oil are convinced that we have very little time to engineer resilience into our communities before the last energy crisis descends. This issue should be of urgent concern to every person who cares about their children, and all who hope there is a viable future for human civilisation post-petroleum."
– Jeremy Leggett, founder of Solarcentury and SolarAid, and author of *The Carbon War* and *Half Gone*

"The Transition movement is the best news there's been for a long time, and this manual is a goldmine of inspiration to get you started."
– Phil England, *New Internationalist*

THE TRANSITION HANDBOOK

From oil dependency to local resilience

by Rob Hopkins
Founder of the Transition Network

We live in an oil-dependent world, and have got to this level of dependency in a very short space of time, using vast reserves of oil in the process without planning for when the supply is not so plentiful. Most people don't want to think about what happens when the oil runs out (or becomes prohibitively expensive), but *The Transition Handbook* shows how the inevitable and profound changes ahead can have a positive effect. They can lead to the rebirth of local communities, which will generate their own fuel, food and housing. They can encourage the development of local currencies, to keep money in the local area. They can unleash a local 'skilling-up', so that people have more control over their lives.

The Transition Handbook is the manual which will guide communities to begin this 'energy descent' journey. The argument that 'small is inevitable' is upbeat and positive, as well as utterly convincing. Read this book!

The Author: Rob Hopkins has long been aware of the implications of our oil-dependent status, and has been energetically campaigning to increase awareness of its impact. Having successfully created an Energy Descent Plan for Kinsale in Ireland which was later adopted as policy by the town council, Rob moved to Totnes in Devon and initiated Transition Town Totnes, the first UK town to address the issues of life after peak oil. He publishes www.transitionculture.org, a popular website that promotes the Transition concept, and www.transitiontowns.org.

ISBN 978 1 900322 18 8   224pp in two colours
£12.95 paperback

THE AGE OF STUPID

> "Captivating and constantly surprising . . . the first successful dramatisation of climate change to reach the big screen."
> – George Monbiot, *The Guardian*

> "Knocks spots off *An Inconvenient Truth*."
> – Mark Anslow, *The Ecologist*

> "Every single person in this country should be forcibly made to watch this film."
> – Ken Livingstone

> "I defy anyone to come out and not feel like they've got to make a difference."
> – Caroline Lucas MEP, Leader of the UK Green Party

> "Can I just pretend I never saw it?"
> – William Nicholson, *Gladiator* writer

The Age of Stupid stars Oscar-nominated Pete Postlethwaite (*In The Name of the Father, Usual Suspects*) as a man living alone in the devastated world of 2055, looking back at old footage from 2008 and asking: why didn't we save ourselves when we had the chance? It was written and directed by Franny Armstrong (*McLibel, Drowned Out*) and exec-produced by John Battsek (Oscar-winning *One Day in September*). The £450,000 budget for the 90-minute film was raised by the filmmaker's innovative 'crowd-funding' model, which gave them complete editorial control over the film's content. 228 ordinary people invested between £500 and £35,000 and each own a percentage of profits. As do the crew, who worked at drastically reduced fees to keep costs down and deliver a big-budget film on the (relatively speaking) tiny budget. A total of 94 tons of CO_2 were emitted during the production of *The Age of Stupid*.

The Age of Stupid has already won a prestigious Grierson Award, and screened at the UN Climate Summit in Poznan in Dec 2008, as well as at the British, EU and Dutch Parliaments. Its UK cinema release is heralded by the world's most inclusive film launch, the People's Premiere – hosted at 64 different cinemas across the nation, including a screening at the Eden Project with an introduction from Rob Hopkins. The People's Premiere is also the world's greenest film launch event, with all power provided by a combination of solar energy and reusing and recycling London's waste, no disposable items, ethically sourced food and drink, and low/zero carbon transport – strictly no flying.

The UK release of the film also launches the social activism campaign, Not Stupid, an ambitious project to turn the film's millions of viewers into climate activists in advance of the UN Climate Summit in Copenhagen at the end of 2009. As well as helping people to understand what 'success' at Copenhagen really means, Not Stupid will also match-make potential climate activists with networks and organizations working towards a low-carbon society and a habitable future for our children, including the Transition movement.